BACKYARD CRITTERS

MARCUS SCHNECK

GALLERY BOOKS
An Imprint of W. H. Smith Publishers Inc.
112 Madison Avenue
New York City 10016

A QUINTET BOOK
produced for
GALLERY BOOKS
An imprint of W. H. Smith Publishers Inc.
112 Madison Avenue
New York, New York 10016

ISBN 0-8317-0669-4

This book was designed and produced by
Quintet Publishing Limited
6 Blundell Street
London N7 9BH

Creative Director: Peter Bridgewater
Art Director: Ian Hunt
Designer: Stuart Walden
Project Editor: Judith Simons
Editor: Henrietta Wilkinson
Picture Researcher: Marcus Schneck
Illustrator: Lorraine Harrison
Artwork: Jenny Millington

Typeset in Great Britain by
Central Southern Typesetters, Eastbourne
Manufactured in Hong Kong by
Regent Publishing Services Limited
Printed in Hong Kong by
Leefung-Asco Printers Limited

Contents

INTRODUCTION 6

SPECIES DIRECTORY 10

INDEX 94

CREDITS 96

Introduction

ABOVE Backyard wildlife brings some problems with it. However, a few tipped garbage cans are a pretty meager price to pay for the entertainment that a raccoon can provide.

"What is a man without the beasts? If all the beasts were gone, men would die from a great loneliness of the spirit. For whatever happens to the beasts soon happens to man.".

Chief Seattle's words ring as true today as when he first spoke them in 1854. Perhaps even truer in the modern context of disappearing and degraded habitat, endangered and threatened species.

It is doubtful that any endangered species will be brought back from the brink of extinction by the habitat provided in a backyard, or even thousands of backyards. Or that the backyards of North America can keep up the pace with which we are destroying open and wild spaces.

But the continent has reached that critical moment when every acre, every quarter-acre, every bit of backyard that we make to return something to the wildlife community is vital.

It does make a difference, and the wildlife will respond. Even balconies and decks can be made to provide some of the essentials of life for a

BELOW Every creature can be a wondrous learning experience, such as the common box turtle that begins many a child's love for wildlife.

OPPOSITE Cloaked by a heavy fog, a white-tailed deer wanders along the edge of a New York backyard.

limited number of species. A moderately sized backyard can provide for even more, a large backyard for a yet larger number. And a wooded backyard with a pond, or a park, can support an almost unbelievable array of animal life.

This is not to say that any one backyard or park can be all things to all creatures. No single spot can attract all of the creatures that are described in this book. No single backyard will play host to every type of creature that is found locally.

But the backyard that is made to provide all the basics of life for wildlife – food, water, sheltered areas and safe spots to raise their next generation – will bring a great diversity into close viewing range.

The backyard "habitater" should not, however, expect to see a lawn filled with the gentle, peacefully co-existing animals that are portrayed in many television nature programs and movies. Nature simply does not always function with a loving, caring hand. Nature functions to allow the fittest of a species to continue the species

into future generations. That means that some die so that others live.

In that context, the backyard wildlife observer will be much better prepared to appreciate the wonders of what he or she is fortunate enough to witness. The miracle of birth – as in the hatching of a robin's egg, or the tiny grass-nest filled with pink and helpless white-footed mouse babies – is an awe-inspiring spectacle. As is the finality of death – as in the devouring of a red eft by a hognosed snake, or the swoop and snatch of a red-tailed hawk taking a song sparrow from the feeder.

And, anyone who goes to the trouble of inviting wildlife into the backyard can expect to see themself, their yard and even their home drawn into this natural web of life, where all things are connected. Woodchucks will dig under the garden fence and eat their way through the choicest vegetable plants. Crickets will invade the home each fall in search of warmth. Birds will beat the property owner to many a ripe fruit or berry.

LEFT The red admiral can be a common backyard visitor in some parts of North America.

LEFT With its cheek pouches packed, the chipmunk is ready to retreat to a storage chamber in its underground burrow.

OPPOSITE The black bear is not yet a common sight in many backyards, but the bruin is extending its range and its proximity to man's habitations.

None of this is either good nor bad. Those are totally human concepts. They are foreign to every other living creature on the face of the Earth. The backyard wildlife observer who leaves them behind upon entering the natural world outside the kitchen window or through the back door will gain a more thorough enjoyment of the experience.

No insect will need to be "nuked" with pesticides. No squirrels will need to be thwarted in their attempts to eat their fill at the bird feeder. And no skunks will need to be trapped out of the area.

Everything will simply exist, going about life in the way that instinct and need tells it to go about life. And from that, the observer will gain the fullest understanding and deepest rewards.

However, the following pages were written with the modern-world situation in mind. We do not propose an idyllic approach to wildlife. While we would not recommend the removal or exclusion of any creature that wants to include the backyard within its territory, we do realize that there are those species that some may not want to go out of their way to attract into the backyard. We have noted the problems that some species can bring with them.

The following pages are far from an all-encompassing look at North America's wildlife. Rather, they are a practical guide to the most common of the continent's critters – mammals, amphibians, reptiles, insects, and birds – that can reasonably be expected to be attracted into the backyard.

Species Directory

Eastern Chipmunk 12
Gray Squirrel 13
Fox Squirrel 14
Red Squirrel 15
Flying Squirrel 16
Woodchuck 17
Thirteen-lined Ground Squirrel 18
Eastern Cottontail 19
Jack Rabbit 20
Masked Shrew 21
Short-tailed Shrew 21
Star-nosed Mole 22
Eastern Mole 22
Little Brown Bat 23
Big Brown Bat 24
Virginia Opossum 25
Harvest Mouse 26
Deer Mouse 26
White-footed Mouse 27
Meadow Vole 27
House Mouse 28
Porcupine 29
Coyote 30
Red Fox 31
Black Bear 32
Raccoon 33
Long-tailed Weasel 34
Eastern Spotted Skunk 35
Striped Skunk 35
White-tailed Deer 36
Eastern Newt 37
Red-backed Salamander 38
Slimy Salamander 38
Tiger Salamander 39
Spotted Salamander 39
Green Frog 40
Bullfrog 40
Wood Frog 41
Pickerel Frog 41
Spring Peeper 42
Common Gray Treefrog 42
Chorus Frog 43
Northern Cricket Frog 43
American Toad 44
Painted Turtle 45
Box Turtle 45
Green Anole 46
Eastern Fence Lizard 46
Five-lined Skink 47
Ringneck Snake 48
Rat Snake 48
Hognose Snake 49
Milk Snake 49
Northern Water Snake 50
Common Garter Snake 50
Smooth Green Snake 51
Bald-faced Hornet 52

Yellow Jacket 52
American Bumble Bee 53
Honey Bee 53
Orange Sulphur 54
Cabbage White 54
Monarch 55
Mourning Cloak 55
American Copper 56
Tiger Swallowtail 56
Silver-spotted Skipper 57
Isabella Tiger Moth 58
Clear-winged Sphinx Moth 58
Cecropia Moth 59
Luna Moth 59
Pennsylvania Firefly 60
Dragonflies & Damselflies 60
Striped Blister Beetle 61
Squash Bug 61
Small Eastern Milkweed Bug 62
Colorado Potato Beetle 62
Thin-legged Wolf Spider 63
Grass Spider 63
Garden Spider 64
Field Cricket 65
True Katydid 65
American Bird Grasshopper 66
Differential Grasshopper 66
Brown Daddy-long-legs 67
Ring-legged Earwig 68
Millipede 68
Ladybug Beetles 69
Japanese Beetle 70
Northern Walkingstick 70
Praying Mantis 71
Little Black Ant 72
Red Ant 72
Earthworm 73
American Robin 74
European Starling 75
Northern Mockingbird 76
White-throated Sparrow 77
House Finch 78
Cardinal 79
American Goldfinch 80
Evening Grosbeak 81
Song Sparrow 82
Barn Swallow 83
House Sparrow 84
Blue Jay 85
White-breasted Nuthatch 86
Black-capped Chickadee 87
Mourning Dove 88
Downy Woodpecker 89
House Wren 90
Common Grackle 91
Red-winged Blackbird 92
Brown-headed Cowbird 93

Eastern Chipmunk

TAMIAS STRIATUS

DATABOX

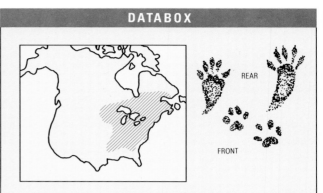

REAR

FRONT

DESCRIPTION: Orange-brown on sides with two side stripes of black; white underside; mottled white; gray, and black on back and top of head; brown tail mottled and edged with black. 5½ to 9½ inches long, with tail 3 to 4½ inches long. Weighs 2¼ to 5 ounces.

SIGN: Nutshells open from one side, scattered on and around flat rock, log, or stump. Small holes in forested slope, which are tunnel entrances. Hindfoot track shows five clawed toes, 1¾ to 3½ inches long; forefoot track shows four clawed toes, about 1 inch long. Walking stride is 7 to 15 inches long, with hindfeet printing ahead of forefeet. Makes a chipping noise.

HABITAT: Wooded and shrubby areas, with ample supply of large rocks and piles of rocks for escape.

REPRODUCTION/REARING: Mates in early spring. Average 3–5 young born in May. Burrow equipped with special birthing/nursery chamber.

HABITS: Makes repeated trips to food sources, stuffing expandable cheek pouches and carrying food off for storage.

HOW TO ATTRACT: Spread nuts and seed on a large, flat, sun-splashed rock that is close to cover.

ONE VARIETY OF CHIPMUNK or another is found in nearly every part of North America, but the Eastern chipmunk is probably the most familiar and widespread. Its range extends across the eastern half of the United States and southern Canada.

The small member of the squirrel family can be found almost anywhere within that range, from the deepest wilderness to the most urban of cities. Wherever its few needs are met – a steady source of its primary food of nuts, seeds, and fruits, and plenty of escape cover – the chipmunk is likely to be found.

True to its Latin genus name (Tamias translates into "storer"), the chipmunk is a horder of food, both temporarily in its super-elastic cheeks and more long-term in its extensive system of multi-chambered underground burrows. It spends much of the winter months in these burrows, but it is not a true hibernator, often emerging on warmer winter days.

Gray Squirrel

SCIURUS CAROLINENSIS

THE MERE MENTION of the gray squirrel is enough to strike panic into the hearts of many a backyard bird watcher. This most common of squirrels has a much-inflated reputation for eating songbird eggs and nestlings, but it is next to impossible to discourage from bird feeders.

However, the bushytail's antics at the backyard feeder more than compensates for the huge amounts of choice seeds that the animal pilfers.

A minute or two on any tree-lined college campus will attest to the ease with which the gray squirrel can be conditioned to beg for food. However, as with all wild creatures, caution must be the watchword, especially when offering food. The animal is unpredictable and can deliver a nasty bite, which can carry the threat of diseases such as rabies.

The gray squirrel generally occurs wherever there are large trees. It will construct its nests both inside hollows in the trees and in the crooks of the larger limbs. But, when the food supply is abundant, the animal will adapt to areas that offer only young, relatively small trees.

When a food supply is depleted, the gray squirrel readily migrates to new "pickings." Records of mass migrations, even across large rivers, have been documented.

DATABOX

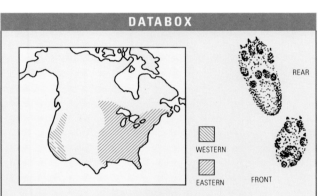

REAR

WESTERN

EASTERN   FRONT

DESCRIPTION: Gray flecked with black and white, and very limited brown that shows more at face and in tail; white underside. Large bushy tail curves in reverse question mark when squirrel sits on hindlegs. 8 to 12 inches long, with 8 to 10 inch tail. Weighs 14 to 25 ounces.

SIGN: Bits of nutshell scattered about forest floor. Small holes well up in tree trunk that have the bark scratched away from repeated entrances and exits. Circular bundles of leaves in branches of trees. Hindfoot track shows five clawed toes, 2¼ to 2¾ inches long; forefoot track shows four clawed toes, 1 inch long. Hindfoot generally prints ahead of forefoot, both in pairs. Walking/ambling stride, 4–6 inches; running stride, up to 3 feet. Tail does not drag. Makes a rasping, barking call at intruders.

HABITAT: Forest areas that include healthy population of nut-bearing trees.

REPRODUCTION/REARING: Mates in winter for litter of 2–3 to be born in spring, and again in late spring for similar litter to be born in late summer.

HABITS: Buries nuts and seeds individually throughout its territory, but never exactly where it has found them.

HOW TO ATTRACT: Any feeder that is not equipped with squirrel-deterrent devices will bring them into the backyard.

Fox Squirrel

SCIURUS NIGER

THIS LARGEST OF OUR TREE SQUIRRELS is regularly misidentified as a big gray squirrel. However it is a distinct species and can easily be identified.

The fox squirrel, which can measure 28 inches from tip of nose to tip of tail, is much larger than the gray squirrel, which reaches a maximum length of no more than 20 inches. In addition, the fox squirrel generally has some yellowish hairs in its tail, while the gray squirrel shows a much more silvery tail.

The two squirrels share much of their range, and also share other characteristics. Both are horders of nuts and seeds, burying them throughout their home territory and later locating some of them by smell (not memory). This habit makes the squirrels an integral part in the spread of nut- and seed-bearing plant species. Many of the tiny parcels are never recovered and soon sprout into new plants.

Fox squirrels are much more likely to choose and repeatedly re-use particular feeding spots within their territories. Large amounts of nut and seed shells can be found by these logs and stumps, and on the ground under these tree branches.

Both species construct the large clusters of leaves in the crooks of tree branches that are commonly known as squirrel nests. However, these serve more as hiding places and resting/sleeping spots rather than as nests for giving birth to the young. Tree hollows more regularly serve that purpose.

DATABOX

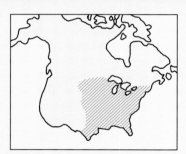

DESCRIPTION: Brown-gray to olive-gray with tan at face and paws; reddish brown underside. Thick bushy tail edged and flecked with reddish brown. Larger than the gray squirrel, 11 to 21 inches long, with 7¾ to 13 inch long tail. Weighs 17½ to 37½ ounces.

SIGN: Similar to gray squirrel, except that nutshell debris is much more concentrated because the fox squirrel prefers to eat at a few selected spots within its territory. Hindfoot track shows five clawed toes, 2¼ to 3½ inches long; forefoot track shows four clawed toes, 1½ inches long. Hindfoot generally prints ahead of forefoot, both in pairs. Walking/ambling stride, 4–8 inches; running stride, up to 3½ feet. Tail does not drag. Makes a rasping, barking call at intruders.

HABITAT: Similar to gray squirrel.

REPRODUCTION/REARING: Mates in midwinter. Litter of 2–4 young are born in late February through early April. Mature females often produce a second litter mid- to late summer. Young stay in nest for 7–8 weeks. Family groups often spend first winter together in den tree.

HABITS: Prefers to carry food to favorite feeding spot to eat.

HOW TO ATTRACT: Any feeder that is not equipped with squirrel-deterrent devices will bring them into the backyard.

Red Squirrel

TAMIASCIURUS HUDSONICUS

ALTHOUGH THE RED SQUIRREL is not uncommon in backyards, it is not nearly as common there as its larger cousins. The smaller (up to 15 inches in length) squirrel's habitat preference of coniferous forests over open deciduous woodland is the reason.

And, while the red squirrel actually inhabits a larger range than the gray and fox squirrels, that range is much more northerly and includes much more land that is not shared with human residents.

This habitat and range preference is based more on food needs rather than any shyness on the part of the red squirrel. The small pine seed and cone-eater is in reality the fiercest fighter among the squirrels. In areas where their ranges overlap, it is not unusual to see the larger gray and fox squirrels being driven off by the smaller fireball.

The red squirrel is also a much noisier squirrel. Its rolling, twirling, chattering call is heard at regular intervals wherever it lives. The comings and goings of other animals can be tracked by these warning calls from the red squirrel.

DATABOX

REAR

FRONT

DESCRIPTION: Red to reddish brown with flecks of gray, more pronounced over shoulders and back of head; white underside, often parted from top of body by distinct black line running from front elbow to rear ankle. Bushy tail of same coloring as body. 7½ to 10 inches long, with 3½ to 7 inch tail. Weighs 5 to 9 ounces.

SIGN: Mounds of cone shells; circular bundles of leaves and shredded bark in coniferous trees. Makes a high-pitched, twirling, chattering call at intruders. Hindfoot track shows five clawed, bony toes, 2¼ to 3½ inches long; forefoot track shows four clawed, bony toes, 1½ inches long. Hindfoot generally prints ahead of forefoot, both in pairs. Walking/ambling stride, 3–7 inches; running stride, up to 30 inches. Tail does not drag.

HABITAT: Coniferous or mixed forests that contain healthy coniferous populations.

REPRODUCTION/REARING: Mates in late winter. Litter of 3–6 born in March to April. Some females produce second litter in late summer.

HABITS: Buries cones similar to the way in which the gray and fox squirrels bury nuts, but in larger caches of many cones each.

HOW TO ATTRACT: Cones gathered when green and later placed at feeding station. As the red squirrel eats bird eggs and baby birds, careful consideration is warranted before trying to attract.

Flying Squirrel

SOUTHERN – GLAUCOMYS VOLANUS; NORTHERN – GLAUCOMYS SABRINUS

RIGHT Flying squirrels do visit feeders in those backyards that border on mature deciduous forests, but leave virtually no evidence of their nighttime presence.

BETWEEN OUR TWO SPECIES of flying squirrel, the entire eastern half of the United States and all of Canada below the timberline falls well within the range of this odd creature. But owing to the totally nocturnal nature of the animal, it is not generally believed to be nearly as common as it is.

The flying squirrel does not actually fly. Instead, it glides from tree to tree on folds of skin that stretch between its front and back legs. A flattened tail, unlike the signature, bushy tails of most squirrels, adds to the animal's aerodynamics. Almost without fail, the flying squirrel lands head-up at its destination.

The flying squirrel is the most carnivorous of the North American squirrels. Nuts, seeds, and vegetable matter make up the bulk of their diet, but bird eggs, baby birds, insects, and other small animals are killed and eaten regularly.

These squirrels do not build leafy nests in the crooks of tree branches. Instead, they generally rely on old woodpecker holes as their homes. They will also use nest boxes, both those intended for squirrel use and those provided for bird nesting.

They do not hibernate, but they will keep to their den trees for extended periods of cold and stormy weather.

DATABOX

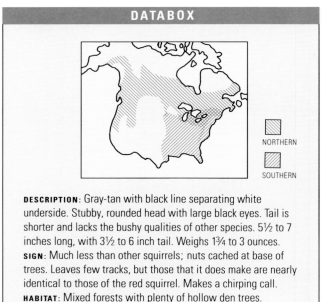

NORTHERN

SOUTHERN

DESCRIPTION: Gray-tan with black line separating white underside. Stubby, rounded head with large black eyes. Tail is shorter and lacks the bushy qualities of other species. 5½ to 7 inches long, with 3½ to 6 inch tail. Weighs 1¾ to 3 ounces.
SIGN: Much less than other squirrels; nuts cached at base of trees. Leaves few tracks, but those that it does make are nearly identical to those of the red squirrel. Makes a chirping call.
HABITAT: Mixed forests with plenty of hollow den trees.
REPRODUCTION/REARING: Mates in late winter. Litter of 2–4 young born about 1½ months after mating.
HABITS: Highly nocturnal, spending days in hollow den trees.
HOW TO ATTRACT: Bird feeder stocked with seeds and nuts if the backyard borders suitable habitat.

Woodchuck

MARMOTA MONAX

FOR AN ANIMAL that musters so very little respect from man, the woodchuck is heavily steeped in folklore. Weather prognostications based on the animal's seeing or not seeing its shadow late each winter are reported – although with tongue in cheek – by the media across the continent.

Also known as the groundhog and the whistle pig, the woodchuck is an animal of the open fields and meadows. However, it has not retreated in the face of man's suburban development. Backyards seem to be totally acceptable to the woodchuck, those with vegetable gardens even more so. One of these animals can devour an entire garden's crop, and fences that hold off other creatures often have little impact on the woodchuck.

In addition, the woodchuck appears to be extending its range into woodlands nearby. This, too, is probably the result of man's opening of areas within the woodlands.

The woodchuck is a true hibernator. Its underground network of runways and rooms – perhaps 50 feet in total length – includes a special, grass-lined hibernation chamber. There it will spend the winter months, its body temperature, breathing, and heart rate all reduced to the barest minimums necessary to keep it alive through the cold period.

Contrary to popular belief, the woodchuck is an excellent tree climber and swimmer.

ABOVE The woodchuck's home range can extend anywhere between two to 160 acres.

DATABOX

FRONT

REAR

DESCRIPTION: Flecked with gray, black, white, and brown; darker at head, with more brown showing at rear of back, rump, and legs. Feet and short, bushy tail are dark brown or black. 16 to 30 inches long, with 3½ to 7 inch tail. Weighs 4¼ to 16 pounds.

SIGN: Large mounds of rock and soil at 8 to 14-inch-diameter entrances to burrow, with distinct trails leading away from the several entrances (all in close proximity to one another). Makes a whistling noise. Hindfoot track shows five clawed toes, 1¾ inches long; forefoot track shows four clawed toes, 2¼ inches long. Hindfoot generally prints ahead of forefoot, both in pairs. Walking/ambling stride, 6–8 inches; running stride, up to 2 feet. Tail does not drag.

HABITAT: Agricultural areas, meadows, regrowth areas, and (more recently) forests.

REPRODUCTION/REARING: Mates in winter. Litter of 4–5 is born in April or May. The young leave the mother's burrow at about 2 months.

HABITS: Stands on hindfeet at main entrance to burrow to survey surrounding area for danger. Hibernates late fall to late winter.

HOW TO ATTRACT: Not a highly desirable species in the backyard, due to extensive tunneling and penchant for eating garden crops.

Thirteen-lined Ground Squirrel

SPERMOPHILUS TRIDECEMLINEATUS

IN THE CENTRAL, plains region of North America, the thirteen-lined ground squirrel is the most common backyard rodent. Man's development of the land seems to agree with the distinctly marked animal, as it is expanding its range in all directions.

Like the much larger woodchuck, the thirteen-lined ground squirrel digs an extensive system of tunnels and chambers. Most of these diggings are less than two feet below the ground's surface, with the exception of the hibernation chamber, which is dug below the freeze line. Also like the woodchuck, this ground squirrel's body temperature, breathing, and heart rate are all extremely reduced during hibernation.

Vegetation and insects make up the bulk of the animal's food preference, but it also kills and eats many species of smaller animals. To the backyard gardener, the thirteen-lined ground squirrel can represent both a threat to crops and an ally in the fight against weed and insect pests.

The animal maintains a near constant surveillance of its territory, often standing on its hindlegs to check for potential threats. When something frightens it, it will dive into the burrow and then emerge just a bit to let out a trilling call.

DATABOX

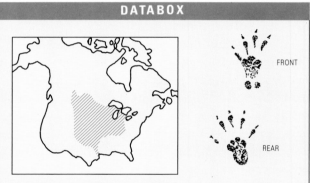

FRONT

REAR

DESCRIPTION: Body is lined with 13 alternating tan and dark brown stripes, front to back; the dark brown stripes carry rows of light tan to white dots. The stripes end at shoulders on sides, but continue onto top of head from back. Face is tan streaked with dark brown and white. 4½ to 7½ inches long, with 2¼ to 5 inch tail. Weighs 3½ to 9½ ounces.

SIGN: Small entrances to burrow with several trails leading away from each; entrances are sometimes hidden. Hindfoot track shows five clawed toes, ¾ inch long; forefoot track shows four clawed toes, ½ inch long. Hindfoot generally prints ahead of forefoot, both in pairs. Walking/ambling stride, 2–7 inches; running stride, up to 14 inches.

HABITAT: Wherever grass remains or is kept short.

REPRODUCTION/REARING: Mates in April. Litter of 7–10 young are born in late May.

HABITS: Similar to woodchuck.

HOW TO ATTRACT: Can be a damaging resident in the backyard, due to tunneling and eating garden crops and ornaments.

Eastern Cottontail

SYLVILAGUS FLORIDANUS

ALTHOUGH THE EASTERN COTTONTAIL'S increasing penchant for nocturnal activity may give the impression that its number is greatly reduced in some locales, the animal remains one of our most abundant backyard wildlife species. Even the most cursory check on the morning of a new snow will offer ample evidence of the animal's presence.

The increased nocturnal nature of the rabbit appears to be a response to heavy pressure from man. The eastern cottontail has mastered the skills of surviving in close proximity to man, despite the constant threat as one of the continent's most popular game species and as a trespasser in man's gardens.

An incredible reproduction rate also works to offset the heavy losses for a species that falls prey to a large array of predators. A single female can produce four litters of as many as nine young rabbits – four to six is more normal – in a given year. A few hours after giving birth, the female will mate once again. The young rabbits nurse for no more than two weeks and are then on their own.

The cottontail's best defense is its ability to remain perfectly still, even when a predator is almost upon it. Failing that tactic, it will dart away at the last minute in a zigzagging path that can be quite confusing to any pursuer.

DATABOX

REAR

FRONT

DESCRIPTION: Mottled and flecked gray and brown, more brown on feet; totally white, cottonball-like tail; white to tan underside; long, erect ears. 14 to 18½ inches long, with 1 to 2¾ inch tail. Weighs 2–5 pounds.

SIGN: Small, low twigs snipped from trees at straight angles. Bark stripped from trunks of small trees. Small, round, black scat in scattered piles along regular trails. The cottontail uses the same passageways through hedgerows and thickets every night, creating clearly visible trails with the vegetation still growing above them. Hindfoot track shows four toes, 3¼ to 4¼ inches long; forefoot track shows four toes, 1½ to 2½ inches long. Hindfoot generally prints ahead of forefoot, in a pair, with the forefeet printing one ahead of the other. Walking/ambling stride, 3–7 inches; running stride, up to 4 feet.

HABITAT: Agricultural areas, forests, backyards, anywhere with ready access to thick, shrubby cover.

REPRODUCTION/REARING: Mates repeatedly, same day that female gives birth, from late winter into fall, producing 4 or 5 litters of 1–9 young each year. Female returns to her nest, which is lined with fur pulled from her belly, to nurse her young each dusk and dawn.

HABITS: Remains totally still until any threat is practically on top of it, then darts off in a zigzagging pattern. Most active dusk and dawn.

HOW TO ATTRACT: Proper habitat is the key. Fresh apples, cut into sections and scattered near trails, are an attractive food offering.

Jack Rabbit

WHITE-TAILED – LEPUS TOWNSENDII; BLACK-TAILED – LEPUS CALIFORNICUS

ONE OF THESE TWO SPECIES of hare (not rabbit) inhabits nearly every bit of suitable range in the western half of the United States. To the north, throughout nearly all of Canada, they are replaced by the snowshore or varying hare (*Lepus americanus*).

The common name of jack rabbit was created by early pioneers because of the resemblance of the hare's long ears to those of the jackass. The ears of a jack rabbit can be as long as 5 inches.

For much of the year, they are most easily differentiated from behind. The black-tailed, as its name implies, has a black area on the top of its tail and on its rump that the white-tailed lacks. In winter, the white-tailed takes on a camouflaging white coat, while the black-tailed remains brown.

Both species are large hares, measuring as much as 26 inches from tip of nose to tip of tail. When undisturbed, they move about in hops of 5 or 6 feet. But, when threatened they can leap up to 20 feet, reach short-burst speeds of 45 miles per hour, and maintain a pace in excess of 30 miles per hour.

DATABOX

REAR

FRONT

WHITE-TAILED

BLACK-TAILED

DESCRIPTION: Flecked gray-tan; white ring around each eye; off-white underside. Hindfeet and ears are enormous in proportion to rest of body. White-tailed jack has white tail and black-tipped ears of same color as rest of body. Black-tailed has tail with black top and black-tipped, nearly naked ears. White-tailed is 22 to 26 inches long, with 2½ to 4½ inch tail; weighs 5½ to 10 pounds. Black-tailed is 18 to 25 inches long, with 2 to 4½ inch tail; weighs 4 to 8 pounds.

SIGN: Regular trails worn through vegetation. Shallow dusting areas. Hindfoot track shows four toes, 2½ to 3½ inches long; forefoot track shows four clawed toes, 1½ inches long. Hindfoot generally prints ahead of forefoot, in a pair, while the forefeet print one ahead of the other. Walking/ambling stride, 9–12 inches; running stride, up to 13 feet.

HABITAT: Grasslands, with thicket areas for cover.

REPRODUCTION/REARING: White-tailed mates in spring. Litter of 2–6 is born a month after mating. They are on their own in another month. Black-tailed mates year-round, producing up to four litters of 2–8 each. After birth, the mother places each of her offspring in separate depressions lined with fur pulled from her belly. The offspring are independent within a month.

HABITS: When pursued the jack rabbit can reach speeds of more than 40 miles per hour over short distances.

HOW TO ATTRACT: Correct habitat.

Masked Shrew

SOREX CINEREUS

ALL SHREWS ARE POPULARLY noted for two things: their diminutive size and enormous appetites. The masked shrew excells at both.

Although it is not the smallest shrew, it never exceeds 5 inches from tip of nose to tip of tail and ½ ounce in weight.

Each day the tiny creature will eat two to three times that weight in snails, slugs, insects, and insect larvae. Its bite carries a poison capable of killing creatures its own size.

The masked shrew, also known as the cinereous shrew, is one of the most widespread of North American mammals.

DATABOX

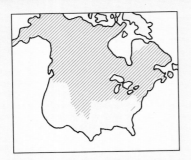

DESCRIPTION: Gray-brown; silver-gray underside; brown, pointed snout is heavily whiskered; tail and feet are brown. 1¾ to 4½ inches long, with 1 to 2 inch tail. Weighs ⅛ to ½ ounce.
SIGN: Leaves very little sign. Sometimes builds leaf nests under logs or in stumps. Hindfoot track shows five toes, ⅛ inch long; forefoot track shows four toes, ⅛ inch long. Hindfoot generally prints ahead of forefoot, both in pairs. Walking/ambling stride, 1–2 inches; running stride, up to 4 inches. Tail does not drag.
HABITAT: Moist fields and moist, open forests.
REPRODUCTION/REARING: Mates spring through fall, producing several litters of as many as 10 young.
HABITS: Eats more than its own weight in insect larvae, worms, and the like, every day.

Short-tailed Shrew

BLARINA BREVICAUDA

THE SHORT-TAILED SHREW is not the largest of our shrews, but it is definitely a contender for the title of most ferocious. Nothing its size or smaller is safe from attack, and much larger animals will receive a painful bite if they disturb the small animal.

A total predator, the short-tailed shrew dispatches its prey quickly with bites to the throat and face. A poison in its saliva instantly immobilizes smaller creatures and can cause pain for several days in humans.

It hunts in underground passageways that have been dug for just that purpose. Males mark their territories with secretions from glands on their undersides that other males respect without fail.

DATABOX

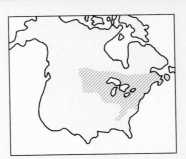

DESCRIPTION: Dark gray, flecked with silver; short, pink snout and pink feet, 3½ to 5½ inches long, with ½ to 1 inch tail. Weighs ½ to 1¼ ounces.
SIGN: Insect shells piled under logs near tunnel entrance of less than 1 inch diameter. Hindfoot track shows five toes, ⅛ inch long; forefoot track shows four toes, ⅛ inch long. Hindfoot generally prints ahead of forefoot, both in pairs. Walking/ ambling stride, 1–2 inches; running stride, up to 4 inches.
HABITAT: Wet, wooded areas.
REPRODUCTION/REARING: Mates in late winter to spring. Litter of 3–5 is born in late spring to summer.
HABITS: A ferocious predator, equipped with venomous saliva.
HOW TO ATTRACT: Correct habitat.

Star-nosed Mole

CONDYLURA CRISTATA

AS THE NAME IMPLIES, the most remarkable feature of the star-nosed mole is a proboscis that ends in a circle of 22 pinkish, tentacle-like projections.

The unusual nose is used in locating prey animals, particularly earthworms, in the soil around its tunnels. When the mole is hunting, the tentacles are in almost constant motion, seeking out the next victim. The star-nosed mole also seeks its prey in the water and is an able swimmer.

Most of the mole's life is spent underground, but males will travel relatively long distances above ground in search of females for mating.

DATABOX

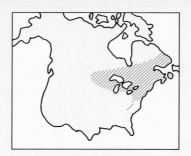

DESCRIPTION: Black over entire body; 22 pink, tentacle-like projections ring the nose; large forelegs are scaly and built for digging. 4½ to 5 inches long, with 2 to 3½ tail. Weighs ¾ to 1½ ounces.

SIGN: Entrances (1½ inch openings) to burrows very close to water. Hindfoot track shows five toes, ¼ inch long; forefoot track shows four toes, ¼ inch long. Hindfoot generally prints ahead of forefoot, both in pairs. Walking/ambling stride, 1–2 inches; running stride, up to 5 inches.

HABITAT: Damp areas across many habitat types.

REPRODUCTION/REARING: Mates in fall. Litter of 3–6 born late winter to mid-spring, and go off on their own after 3 weeks.

HABITS: An able swimmer, relying on food obtained underwater when the ground is frozen.

HOW TO ATTRACT: Correct habitat.

Eastern Mole

SCALOPUS AQUATICUS

LIKE MOST MOLE SPECIES, the eastern (or common) mole is generally less likely to be detected by sightings of the animal itself than by the raised ridges that its tunnels push up through the lawn, and by its raised molehills of excavated soil.

Throughout most of the year, encounters between males will generally result in struggles to the death. But in the late winter/early spring mating season, several males might be found in the same tunnel.

DATABOX

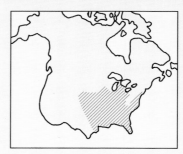

DESCRIPTION: Gray to brown across entire body. Forefeet are wider than they are long, heavily clawed, with palms turned to the outside of the body. Short, naked tail. Body, 3–9 inches long, with a ¾ inch tail. Weighs 3–6 ounces.

SIGN: Evidence of mole tunnels beneath the surface of the lawn.

HABITAT: Weedy and grassy areas, with loose, dry soil.

REPRODUCTION/REARING: Mates in late winter. Female bears one litter of 2–6 in early spring in a special maternity chamber of her burrow, lined with soft grasses. The young are on their own at about 1 month.

HABITS: Almost totally subterranean.

HOW TO ATTRACT: Correct habitat.

Little Brown Bat

MYOTIS LUCIFUGUS

OF THE FORTY OR SO species of bat across North America, the little brown bat is by far the most common. It can be seen in virtually all environments where it can find a nightly supply of its insect prey.

This is the species of bat most commonly found in homes and quickly dispatched with a broom. However, such killing is totally unnecessary. An old shirt or jacket thrown over the bat, when it comes to rest, and quickly wrapped around it will allow for the animal's removal to the outdoors. Leave the bat alone and it will untangle itself from the shirt and fly free. Care must always be exercised in close contact with any bat as all species are potential rabies carriers.

Bats are the only truly flying mammal. Other remarkable tales about the small creatures – such as becoming tangled in human hair and sucking blood – are generally not totally factual for most North American species. A few tropical species do subsist on blood.

DATABOX

DESCRIPTION: Shiny brown in mottled shades; cream underside. Ears are more than ½ inch long. Body, 3 to 3¼ inches long, with 1 to 1½ inch tail. Weighs ⅛ to ¾ ounce.
SIGN: Scat beneath daytime roosting spot.
HABITAT: Varied; near good supply of night-flying insects.
REPRODUCTION/REARING: Mates in fall, but fertilization delayed until following spring. A single baby is born in spring. It nurses for 3 weeks and then flies on its own.
HABITS: Hibernates early fall through late winter.

LEFT The little brown bat, at approximately 3 inches long, is in fact not much smaller than the big brown bat which grows to 5 inches. This little brown bat was discovered in an attic and photographed devouring an insect.

Big Brown Bat

EPTESICUS FUSCUS

ALTHOUGH BIG BROWN BATS are occasionally seen in flight during the day, they generally emerge from their daytime shelter at dusk to hunt for their insect prey.

Using echolocation – emitting high-pitched sounds that bounce off objects and return as echoes – the bats avoid objects and find the insects in the darkened skies over country and city alike.

Although they generally are not found in great numbers in the cities because of the relative shortage of insects, bats are commonly seen snatching insects under city street lights.

DATABOX

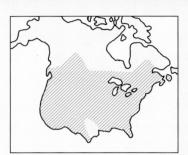

DESCRIPTION: Varying shades of brown; tan underside. Wings are without fur. 4 to 5 inches long, with 1¾ to 2¼ inch tail.
SIGN: Scat beneath daytime roosting spot.
HABITAT: Varied, but near abundant supply of night-flying insects.
REPRODUCTION/REARING: Mates in fall, but fertilization delayed until following spring. A single baby is born in spring. It is independent after 3 weeks.
HABITS: Hibernates fall through late winter.

RIGHT This big brown bat was caught by the camera resting on a piece of wood. The tan colored fur of the belly, lighter than the varied brown shade of the head and back, is clearly visible.

Virginia Opossum

DIDELPHIS VIRGINIANA

THE VIRGINIA OPOSSUM, more commonly known as the possum, is North America's only marsupial. This is an order of mammals in which the female lacks a placenta and instead carries her young in a marsupium, a pouch on her belly.

As such, the possum has an extremely short gestation period for a mammal. After less than two weeks inside their mother, 1–12 bumble bee-sized babies emerge and crawl along a channel on the female's belly to her pouch. Inside the pouch, each one attaches itself to a nipple. There it will stay for the next two to three months, when it will begin to venture outside to ride on its mother's back.

Marsupials are extremely ancient mammals. The possum, in fact, is North America's most ancient mammal, existing in a form close to what it is today about 60 million years ago.

Versatility is key to such a long existence on Earth. The possum will eat almost anything organic. It will quickly become a regular visitor to garbage cans and any discarded wastes. Such versatility, combined with the ample discards of human society, is also allowing the animal to extend its range northward.

When threatened, the possum will hunker down and hiss, showing its set of sharp teeth. If that fails, it will "play dead," flopping onto its side, closing its eyes, letting its stuck hang from the side of the mouth and remaining quite still. Finally, if all else fails, it will release a foul-smelling substance from a gland beneath the tail.

DATABOX

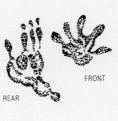

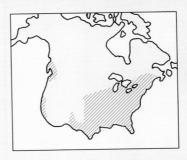

REAR   FRONT

DESCRIPTION: Greasy appearance with long, thin gray to white hairs over a darker fur; silver-gray, pointed face with naked pink to black/pink-tipped ears; naked, scaley tail. 25 to 42 inches long, with 10 to 20 inch tail. Weighs 4 to 16 pounds.

SIGN: Hindfoot track shows five clawed toes, with thumb-like appearance on inside, 2½ inches long; forefoot track shows five toes, 1¾ inches long. Hindfoot generally prints next to forefoot, both sides in pairs. Walking/ambling stride, 7–10 inches; running stride, up to 15 inches.

HABITAT: Open forests, agricultural areas, suburbia.

REPRODUCTION/REARING: Mates in late winter. Litter of up to 20 young is born 12–14 days later. Each baby crawls to brood sac and attaches itself to 1 of mother's 13 teats. Late arrivals that cannot find a teat die very shortly. At 5–6 weeks the young may leave the pouch temporarily to crawl on the mother, but continue to nurse for 2 months until they go off on their own.

HABITS: Makes regular visits to any large source of carrion. In extremely cold periods, it will stay in the den until it can stand the hunger no longer.

HOW TO ATTRACT: Food scraps.

Harvest Mouse

EASTERN – *REITHRODONTOMYS HUMULIS*
WESTERN – *REITHRODONTOMYS MEGALOTIS*

THE FIVE NORTH AMERICAN SPECIES of the harvest mouse are horders of seeds, which they store in their round, grass-lined nests as well as in nearby caches. In the summer months, they will also eat a great many insects and new plant shoots.

All five species closely resemble the non-native house mouse, which is much more of a pest to man. Close inspection of the rodent's upper incisors is a reliable means for differentiating the harvest from the house: each of the harvest mouse's incisors will have a groove from top to bottom.

DATABOX

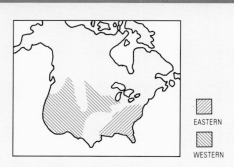

EASTERN
WESTERN

DESCRIPTION: Dark brown to gray-brown; white to buff underside. 4¼ to 6¾ inches long, with 1¾ to 3¾ inch tail. Weighs ¼ to ¾ ounces.
SIGN: Builds round, baseball-sized nest of grass, above ground, attached to weed stalks. Hindfoot track shows five toes, ¾ inch long; forefoot track shows four toes, ¼ inch long. Hindfoot generally prints ahead of forefoot, both in pairs. Walking/ambling stride, 2½ inches; running stride, up to 6 inches.
HABITAT: Agricultural and new growth areas with plenty of shrubby undergrowth.
REPRODUCTION/REARING: Mates year-round. Female gives birth to 2–7 young 3 weeks after mating. Weaned in 2–3 weeks.
HABITS: Caches seeds and other food matter in nests and nearby.
HOW TO ATTRACT: Habitat.

Deer Mouse

PEROMYSCUS MANICULATUS

THIS IS THE SPECIES OF MOUSE most commonly found in homes about the time of the first frost each fall. The small rodents will occupy every niche in their search for warmer shelter as the nights grow colder. A highly social animal, a dozen or more might gather in such sheltered spots for the winter.

Their more natural homes are in brush piles and wood piles, where their nest is a ball of grass, feathers, twigs, string, and whatever other short, dry substance they can find and collect.

DATABOX

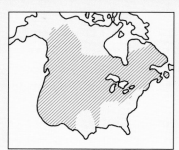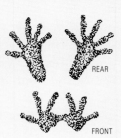

REAR

FRONT

DESCRIPTION: Dark gray-brown to red-brown; white underside; pinkish feet. 4½ to 8 inches long, with 1¾ to 4 inch tail. Weighs ¼ to 1¼ ounces.
SIGN: Regularly used trails. Hindfoot track shows five toes, ¾ inch long; forefoot track shows four toes, ¼ inch long. Hindfoot generally prints ahead of forefoot, both in pairs. Walking/ambling stride, 2½ inches; running stride, up to 6 inches. Tail drags in snow.
HABITAT: Brushy and wooded areas.
REPRODUCTION/REARING: Mates year-round, producing 3–4 litters per year. Litter of 2–6 is born 3 weeks after mating. They are weaned 2–3 weeks later.
HABITS: Digs several burrows, for regular use and for escape.
HOW TO ATTRACT: Seeds on ground, near thickets, brush piles and wood piles.

White-footed Mouse

PEROMYSCUS LEUCOPUS

THE WHITE-FOOTED MOUSE is an extremely adaptable, and consequently widespread, species. It prefers open forest floors and brush.

Nests of grass and other plant fibers are built in log piles, brush piles, hollow logs, bird nests, and stumps. Each nest is used only until it becomes contaminated with scat; the occupant then moves to a new location.

The white-footed mouse is active year-round, but the colder winter days are generally spent asleep.

DATABOX

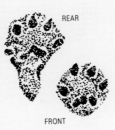

DESCRIPTION: Gray-brown to orange-brown; white to buff underside; pinkish feet; ears are ½ inch long. Body, 6 to 8¼ inches long, with 2½ to 3½ inch tail. Weighs ½ to 1½ ounces.

SIGN: Fruit pits and stones, particularly cherry, cached under logs and in stumps. Hindfoot track shows five toes, ¾ inch long; forefoot track shows four toes, ¼ inch long. Hindfoot generally prints ahead of forefoot, both in pairs. Walking/ambling stride, 2½ inches; running stride, up to 6 inches. Tail drags in snow.

HABITAT: Shrubby and forested areas.

REPRODUCTION/REARING: Mates year-round, producing 3–4 litters per year. Litter of 2–6 is born 3 weeks after mating. They are weaned 2–3 weeks later.

HABITS: Remains in nest on particularly cold winter days.

HOW TO ATTRACT: Cherry pits near thicket, brush or wood pile.

Meadow Vole

MICROTUS PENNSYLVANICUS

MORE COMMONLY KNOWN as the meadow mouse or field mouse, this rodent is a strong contender for the title of most numerous U.S. mammal. It regularly occurs in populations of hundreds to an acre, with "epidemics" of up to 10,000 to an acre.

In such numbers, and with each female capable of producing more than 100 young mice each year, the meadow vole can cause significant damage to grain crops. However, it is also a staple in the diet of many predators.

In the backyard, even a few square yards of unmown lawn can become home to the meadow vole's ground-level system of tunnels, bedding, and feeding areas.

DATABOX

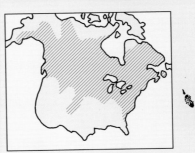
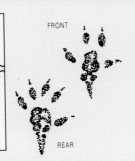

DESCRIPTION: Shades of brown flecked with black and white; silver-gray underside. Tail is dark brown above, lighter below. Feet are dark brown. 5¼ to 7¾ inches long, with 1¼ to 2½ inch tail.

SIGN: Trails about 1½ inches across running through thick weeds and grass, with small piles of cut grass blades along trails. Hindfoot track shows five toes, ¾ inch long; forefoot track shows four toes, ¼ inch long. Hindfoot generally prints ahead of forefoot, both in pairs. Walking/ambling stride, 2½ inches; running stride, up to 6 inches. Tail drags in snow.

HABITAT: Grassy, weedy areas.

REPRODUCTION/REARING: Mates year-round. Litter of 4–8 born 3 weeks after mating. Female mates again, almost immediately. One female can produce 13 litters per year. Young are weaned in 10–12 days.

HABITS: When threatened, the meadow vole pounds its hindfeet on the ground.

HOW TO ATTRACT: Allow corners of the backyard to go unmown.

House Mouse

MUS MUSCULUS

THE HOUSE MOUSE originated in Asia and quickly spread across the globe. It came to North American shores as early as the 16th century, on board the ships of European explorers.

Living in loosely knit, highly migratory colonies, house mice are able to take advantage of even the most temporary niches, such as a farmer's grain field. They'll move in, produce several litters, and then move on when the field is harvested.

A penchant for chewing anything chewable that it encounters makes the house mouse an extremely destructive invader in both country and city.

ABOVE A house mouse hastily exits from a hole gnawed in the base of a wooden wall.

DATABOX

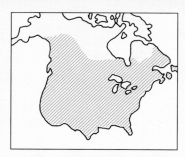

DESCRIPTION: Gray-brown, sometimes lighter on underside; naked tail. 5 to 7¾ inches long, with 2¼ to 4 inch tail. Weighs ½ to 1 ounce.

SIGN: Small holes in walls. Small pellet-like scat. Shredded materials of all sorts. Hindfoot track shows five toes, ¾ inch long; forefoot track shows four toes, ¼ inch long. Hindfoot generally prints ahead of forefoot, both in pairs. Walking/ambling stride, 2½ inches; running stride, up to 6 inches.

HABITAT: Near man.

REPRODUCTION/REARING: Mates year-round, producing litters of 3–16 about 3 weeks after mating. Young are weaned in another 3 weeks. Populations tend to boom and crash, relocating with local food supplies.

HABITS: Lives in colonies; each mouse maintains its own nest.

HOW TO ATTRACT: Not a desirable species.

Porcupine

ERETHIZON DORSATUM

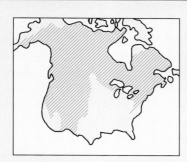

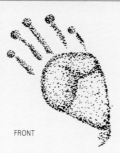

FRONT

DESCRIPTION: Black or dark brown to yellow-brown. Dark, non-quilled face. Long gray-white quills cover rest of body. 25 to 37 inches long, with 5½ to 12 inch tail. Weighs 7½ to 40 pounds.

SIGN: Bark stripped from tree trunks and limbs, with noticeable teeth marks into the remaining wood. Large accummulations of pellet-like scat at entrances to dens in rocks. Hindfoot track shows five clawed toes, 2¼ inches long; forefoot track shows five toes, 1¾ inch long. Hindfoot generally prints ahead of forefoot, both sides in pairs. Walking/ambling stride, 6–10 inches; running stride, up to 24 inches.

HABITAT: Heavily forested areas.

REPRODUCTION/REARING: Mates October to November. Single offspring born 209–220 days later. The baby can follow its mother through the tree limbs within 2 days and is weaned at 4 months.

HABITS: A very slow-moving, docile animal. Constantly in search of salt.

HOW TO ATTRACT: Salt blocks placed in a backyard that is adjacent to heavily wooded forest. However, once attracted, it can cause a great deal of damage through its bark stripping and lust for salt.

WHILE THE PORCUPINE is an interesting study in animal defenses, given its armory of barbed-pointed quills and its slow, meandering manner, it can also be an annoying nuisance when living close to man.

Those quills are a constant, painful threat to man's domestic animals, although one encounter is generally enough to put a dog off of "porkies" for life. The quills cannot be "shot" at an attacker, but they do pull free of the porcupine quite easily on contact and they are painful to remove.

The animal's wintertime girdling of trees for the inner bark leaves a wake of stunted and dying trees. And its constant search for salt, including the salt in human sweat, leads the large rodent to gnaw at many of man's objects.

The porcupine much prefers the deep woodlands, particularly coniferous forests. But, as man extends his developments into these areas, and the porcupine extends its range to the south, encounters are growing more frequent. The "porkie," easily discharged by man's devices, is generally the loser.

REAR

Coyote

CANIS LATRANS

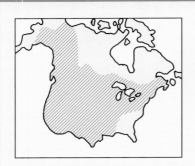

REAR

DESCRIPTION: Salt-and-pepper gray or reddish gray above with buff underside. A dark, vertical line extends down the lower portion of each foreleg. Bushy tail with black tip; held down when running. Prominent ears. Stands 23–27 inches at the shoulder and is 40–52 inches long, with an 11–15 inch tail. Weighs 20–50 pounds.

SIGN: Doglike scat, often full of hair. Tends to use regular hunting trails that are long, meandering and scat-littered. Track shows four toes, all with claws, about 3 inches long. Tracks generally appear in a straight line. Walking stride, 13 inches; trotting, 24 inches; running, 30 inches; can leap 14 feet. Often howl to one another at night.

HABITAT: Open plains and brushy areas in the West; brushy areas in the East.

REPRODUCTION/REARING: Mates January through April; average 6–8 pups (litters of 1 to 19 have been recorded) born 60–65 days later. Dens wherever hole-like protection can be found. Young are nursed for 2 weeks, venture out on their own at 6 weeks. Family remains together at least until fall.

HABITS: Wary of man, elusive. Most active at dusk, night, and dawn. Caches uneaten portions of food by burying them.

HOW TO ATTRACT: Best not to offer food; secure garbage cans.

A MASTER OF ADAPTATION, and a total opportunist in feeding habits, the coyote has claimed nearly all of North America as its own in this century.

Despite the best efforts of man to eradicate the large canine – including massive, public-funded poisoning, trapping, and hunting programs – the animal continues to increase its numbers and its range. The only effect that large-scale anti-coyote efforts appear to have is triggering a baby boom among the remaining population.

The East was the last non-coyote stronghold until quite recently. But, never one to let suitable and even marginally suitable habitat go to waste, the coyote has colonized the region quickly. A huge niche was left for a large predator as man drove out and killed off those creatures less able to adapt, such as the timber wolf and Eastern mountain lion. Taking advantage of the void,

the coyote migrated along an arc to the north and east of the Great Lakes into Ontario and then south into the eastern United States. Along the way, it apparently interbred with wolves and produced a significantly larger version of itself. A male coyote in New Mexico averages 24 pounds; in Vermont, 39 pounds.

The coyote is also extending its range into the suburban and urban regions of man. In October 1987, coyotes killed 48 flamingos in the Los Angeles zoo. Since 1974, there have been 16 recorded attacks on humans in Los Angeles County, one that resulted in the death of a 3-year-old girl just outside her home. Road-killed coyotes have been found within the city limits of Pennsylvania's capital city, Harrisburg. During the severe winter of 1988, animal control officers in Republic, Mo., were trying to kill a few of the animals that had moved into town to scrounge from garbage cans.

Because of their opportunistic nature, the actions of a few coyotes generally earn the animal a bad reputation wherever it lives. Domestic sheep are basically dumb animals, and a coyote that learns this will quickly take to raiding flocks. Garbage cans offer even easier meals, but when the human owner becomes irritated with the mess of dumped garbage cans and removes the source of food, the coyote will look for other sources in its now established territory: domestic pets and even humans are likely targets. In some areas, laws have been passed to ban such feeding of coyotes. In all areas, feeding of any large predator is ill-advised.

Red Fox

VULPES VULPES

THE RED FOX carries a well-earned reputation for cunning and stealth. It has learned to exist in the shadows of man's developments, even while man sought to destroy it as a pilferer of domestic animals or as a sport animal.

Backyards in the vicinity of abandoned fields, open forests, or brush-covered vacant lots – virtually anywhere on the continent – may be included in the animal's nighttime hunting grounds without the owners ever knowing of the fox's proximity.

On the other hand, individuals and families have been attracted to make regular appearances during daylight hours and at close range by man's discards, notably fish heads and entrails.

Dawn and dusk remain the optimum times for observation, for it is then that the animal is most active. But, on sunny winter days, the red fox can be seen napping on exposed banks of snow with its bushy tail wrapped across its nose.

The red fox is a highly territorial animal, maintaining a system of regular "scent post" rocks, weed stems and the like, on which it has urinated, throughout its home range. However, it will quickly abandon that territory if under constant threat from man or other animals.

DATABOX

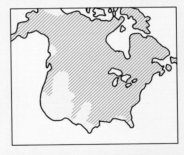

FRONT

REAR

DESCRIPTION: Red to red-brown; white underside continuing up throat and chin. Bushy tail of same color as body ends in white tip. Feet, lower legs, and backs of ears are black. Stands 14 to 17 inches at shoulder and is 35 to 44 inches long, with a 13 to 18 inch tail. Weighs 7 to 16 pounds.

SIGN: Den with well-excavated entrance hole usually made from the den of some other animal, such as the woodchuck or marmot. Food is cached nearby. Hindfoot track shows four clawed toes, 2¼ inches long; forefoot track shows four clawed toes, 2½ inches long. Forefoot prints ahead of hindfoot, in zigzagging line. Walking/ambling stride, 8–14 inches; running stride, up to 36 inches.

HABITAT: Agricultural and open wooded areas.

REPRODUCTION/REARING: Mates January to March. Litter of 4–10 kits is born 51 days after mating. Kits are weaned at about 1 month and begin playing outside near the entrance of the den, eating prey that is brought to them by the parents (regurgitated meat at first, but soon live). Leave the den site at 5 months.

HABITS: An extremely elusive and nocturnal animal. In winter may be found asleep on snow bank with tail curled around nose.

HOW TO ATTRACT: Fish heads and discards may bring the red fox into close range.

Black Bear

URSUS AMERICANUS

MANY GUIDES TO BACKYARD wildlife do not include the black bear. But, as the bruin's range continues to expand to the south and west, there can be no doubt that more and more of the big black creatures are regularly turning up in backyards.

Reports of black bear being relocated after wandering into highly urbanized areas are becoming almost commonplace. A sizeable number of individuals have taken to baiting nightly bruin visitors with temptations such as day-old doughnuts and sweet corn. In some parts of the bear's range, individual animals have taken to denning under back porches.

The black bear is definitely becoming at least a sometimes backyard creature. However, such close encounters with an animal that can weigh several hundred pounds and is able to shred large logs with a swipe of its claws are best not encouraged.

Unprovoked attacks on humans are extremely rare. Some would say nonexistent in recent decades. On the other hand, few of us are knowledgeable enough in the ways of the bear to recognize every little thing that might provoke the animal. At the very least, such a large animal possesses a great deal of potential for property damage in pursuit of any food items it wants.

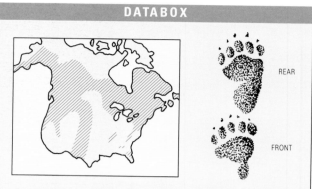

REAR

FRONT

DATABOX

DESCRIPTION: Black to cinnamon, sometimes with white patch on center of chest; only other coloring is a tan snout. Stands 3 to 4½ feet at shoulder and is 4½ to 7 feet long, with a 3 to 6 inch tail. Weighs 200 to 700 pounds.

SIGN: Logs and stumps ripped apart and turned over in search for insects. Trees marked with vertical claw scatches and bites. Dog-like, but larger, scat filled with insect parts, partially digested vegetation, and seeds. Well-worn trails leading to swamps and bogs. Hindfoot track shows five clawed toes, up to 7 inches long; forefoot track shows five clawed toes, up to 4¾ inches long. Hindfoot prints nearly straight ahead of forefoot, both sides in pairs. Walking/ambling stride, up to 24 inches; running stride, up to 60 inches. Coarse black hairs on trees where bear rubbed its backside.

HABITAT: Wooded and wetland areas.

REPRODUCTION/REARING: Mates June to July. 1–5 cubs born January to February, while female is in hibernation. Most females mate every other year. Their cubs often stay with them through a second year's hibernation.

HABITS: Constantly in search of its diverse food supply. Can cause damage to man's beehives, orchards and even buildings as part of this search.

HOW TO ATTRACT: Although unprovoked black bear attacks are extremely rare, or even non-existent, it is generally not a good idea to desensitize such a large, powerful animal from its natural wariness of man.

RIGHT A large black bear in its natural woodland habitat.

Raccoon

PROCYON LOTOR

THE RACCOON has done quite well in all of man's environments. It is an extremely adaptable animal, with agile front paws that it can use nearly as effectively as hands. Almost every food situation that man's existence offers is an opportunity for the raccoon, no matter how well secured.

Garbage cans are a favored target, and in this respect the raccoon is not a tidy eater. Objects that cannot be eaten are scattered and shredded as the raider searches for tastier morsels.

Any spot that proves worthwhile in terms of food found will become part of the raccoon's nightly routine. Some are so regular in their rounds of such locations as to appear within minutes of the same time every evening, so long as the food supply continues.

However, serious consideration must be given before any effort is made to attract raccoons. They are a principal carrier for diseases such as rabies and distemper. In addition, they are effective predators on any animals that are smaller than themselves.

They do show a fondness for water, but do not insist on washing every morsel of food before eating it. This myth probably arose because raccoons do much of their hunting by groping among the pebbles and gravel of shallow water with their front paws. Some theories have also been advanced that the water makes swallowing easier for the animals.

DATABOX

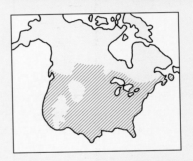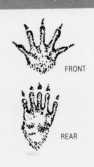

DESCRIPTION: Gray to gray-brown or red-brown, 5 to 9 alternating rings of black and gray-brown around the bushy tail, and a black mask outlined in silver-gray across the eyes. 23 to 38 inches long, with a 7 to 16 inch tail. Weighs 10 to 48 pounds.

SIGN: Cylindrical scat on logs and rocks near streams. Cracked crayfish shells along waterways. Hollow den tree with bark rubbed and clawed away from entrance hole, and much scat around the base of the tree. Hindfoot track shows five clawed toes, 3¾ inches long; forefoot track shows five clawed toes, 3 inches long. Hindfoot prints behind and inside forefoot, both sides in pairs. Walking/ambling stride, 12–18 inches; running stride, up to 30 inches.

HABITAT: Wooded and shrubby areas near streams, rivers, lakes, and ponds.

REPRODUCTION/REARING: Mates in February to March. Litter of 3–7 is born April to May. The eyes of the young remain shut for about 3 weeks. By 8 weeks they are weaned and playing at the den entrance. Some go off on their own soon thereafter, while others stay with the mother until the following spring.

HABITS: Most active from dusk to dawn. A great deal of time is spent in and along waterways. Good nut and berry crops will draw the animal quite a distance inland.

HOW TO ATTRACT: Caution is advised; raccoons are a primary species involved in the rabies situation, but eggs and sweets, such as honey, are favorites.

Long-tailed Weasel

MUSTELA FRENATA

RIGHT Displaying its all-white winter coloring, only the head of this long-tailed weasel is visible above the camouflaging snow.

ANOTHER POTENTIAL UNDETECTED VISITOR to the backyard might be the long-tailed weasel. The silent, deadly predator will readily take advantage of the rodents attracted to a backyard, particularly if that backyard is near wooded areas with plenty of brushy cover. Meadow voles are a favored prey species.

The weasel is relentless in its pursuit of prey, attacking animals much bigger than itself and overcoming substantial obstacles to get to its victim.

The word "bloodthirsty" has been applied to the weasel, sometimes justifiably. The blood of a kill can trigger the killing instinct in the animal, leading to attacks on all other potential prey that is immediately available. This has earned the animal a localized reputation as a wanton killer of entire flocks of farmer's chickens. Bounties have been offered on this smallest of all carnivores.

A relative of the skunk, the weasel is equipped with anal scent glands that can release a foul-smelling, musky substance when the animal is threatened or excited. It appears to use this same liquid in communicating with others of its species.

The long-tailed weasel is one of those select species that turn white in the winter, although this trait appears to be genetically confined to those populations in the northern part of the animal's range.

DATABOX

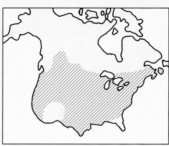

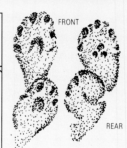

FRONT

REAR

DESCRIPTION: Shades of brown with limited and scattered white flecking; white underside through throat to chin. All white during the winter months in the northern part of its range. 11 to 22 inches long, with 3 to 6½ inch tail. Weighs 3 to 9½ ounces.

SIGN: Dead rodents cached under log. Small, but long scat containing small pieces of bone on rocks and logs. Hindfoot track shows five toes, 1½ inches long; forefoot track shows five toes, 1¼ inches long. Hindfoot prints almost directly behind forefoot, with overlap, both sides in pairs. Walking/ambling stride, 6–10 inches; running stride, up to 14 inches.

HABITAT: Woodlands and agricultural areas.

REPRODUCTION/REARING: Mates in midsummer. Litter of 3–9 young is born following April to May. The offspring go off on their own at about 2 months.

HABITS: Extremely efficient at hiding all comings and goings from rest of world. Will regularly use a hedgerow, stone wall, or similar cover to mask its daily movements.

HOW TO ATTRACT: Healthy rodent population.

Eastern Spotted Skunk

SPILOGALE PUTORIUS

ALTHOUGH THE STRIPED SKUNK is the image that comes more immediately to mind at the mention of the generic family name, the eastern spotted skunk actually carries a more potent defensive spray.

It also exhibits a more unusual warning of its readiness to unleash that spray. When frightened, the spotted skunk does a handstand on its front paws, with its rump pointed at the aggressor.

DATABOX

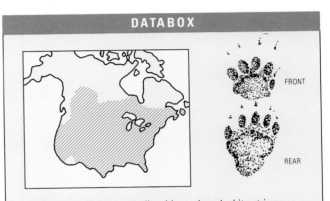

DESCRIPTION: Black with broad, irregular bands of white spots; bushy, black tail with white tip. Body, 13½ to 23 inches long, with 2¾ to 9 inch tail. Weighs 1¾ to 2¼ pounds.

SIGN: Small mounds of earth excavated when digging for insects. Hindfoot track shows five clawed toes, 1¼ inches long; forefoot track shows five clawed toes, 1 inch long. Hindfoot prints behind forefoot with slight overlap, both sides in pairs. Walking/ambling stride, 6–8 inches; running stride, up to 14 inches.

HABITAT: Open woodlands and agricultural areas.

REPRODUCTION/REARING: Mates in late winter. Litter of 2–10 young is born in mid-spring.

HABITS: Most active from dusk to dawn. Travels random nighttime route that will return to areas of previous food finds.

HOW TO ATTRACT: Table scraps and cat food, but caution is advised because of the animal's foul-smelling, long-lasting spray and its involvement with rabies.

Striped Skunk

MEPHITIS MEPHITIS

WHEN THREATENED, the striped skunk stamps its front feet, hisses and raises its bushy tail in warning. If the threat passes, the act will go no further. But, if the aggressor persists, it will be met by a foul-smelling, long-lasting spray from the skunk's anal glands, a spray that can reach up to 20 feet away.

Skunks are opportunistic feeders and will quickly respond to any food that is offered to them. Caution is advised, however, because a colony of skunks is quick to establish and slow to disburse. And, wherever there are skunks there is the constant threat of skunk spray, as well as diseases such as rabies.

DATABOX

DESCRIPTION: Black, generally with two broad white stripes along back, but white coloration is extremely variable. Bushy black tail generally has white edges and/or tip. 20 to 33 inches long, with 7 to 16 inch tail. Weighs 5 to 13 pounds.

SIGN: Small mounds of earth excavated when digging for insects. Hindfoot track shows five clawed toes, 1½ inches long; forefoot track shows five clawed toes, ¾ inch long. Hindfoot prints almost directly behind forefoot, both sides in pairs. Walking/ambling stride, 6–10 inches; running stride, 12–18 inches.

HABITAT: Open woodlands and agricultural areas.

REPRODUCTION/REARING: Mates in February to March. Litter of 3–7 is born 62–64 days later. The young are weaned in 6–7 weeks.

HABITS: See spotted skunk.

HOW TO ATTRACT: See spotted skunk.

White-tailed Deer

ODOCOILEUS VIRGINIANUS

THE WHITE-TAILED DEER is another species that is experiencing a population explosion in the shadow of man and consequently coming to inhabit areas in close proximity to its benefactor. However, it generally remains an elusive creature, observed mostly at dusk.

Some of the increase must be attributed to the demise of the deer's natural predators, such as the wolf and mountain lion, over much of its range. But, much of the population upswing – from near extinction in the Northeast and Midwest at one time – is artificially-enhanced by game agencies because of the animal's popularity as a big game species for hunters. A side-

DATABOX

 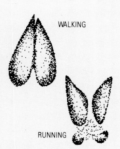

WALKING

RUNNING

DESCRIPTION: Red-brown to gray-brown; white underside through throat to chin, some in face and insides of large ears. Large, flag-like tail is brown with white edges on top and white underneath. Bucks and some does have forward-curving antlers. Stands 3–4 feet at shoulder and is 4½ to 6 feet long, with a 6–14 inch tail. Weighs 100 to 300 pounds.

SIGN: Regular trails from bedding areas to feeding areas. Pressed down leaves and litter where the deer was bedded down. Large, black, pellet-like scat. Bark rubbed from trees in "buck rubs." Leaves scraped away, generally under a small tree in which the branches have been thrashed by antlers. Hindfoot and forefoot are each two separate, slightly curved hoof segments, 2 to 3¼ inches long. Front and hind print in nearly straight line. Hoof segments will spread toward outside and dewclaws may show at rear of track when running. Walking/ambling stride, 18–24 inches; running stride, up to 10 feet.

REPRODUCTION/REARING: Mates in late fall. Regularly twins, but also single offspring, triplets and quadruplets are born 7 months after mating. Fawns follow mother at 4–5 weeks and are weaned at 4 months. They remain with mother at least through the following winter.

HABITS: Most active dusk to dawn. Takes advantage of any available agricultural fields and orchards.

HOW TO ATTRACT: Apples and salt blocks placed in backyards that are isolated and undisturbed enough to put the animals at ease.

effect is deer populations too large for their small wooded areas and pockets that are too close to human habitations to allow hunting.

This situation is pushing more and more of the animals into backyard environments. They bring with them the thrills of a truly large wild animal, as well as many problems such as increased auto-deer collisions.

White-tailed deer can be conditioned to reappear in the same general area each evening through a regular offering of food. Apples are a favorite. A block of salt, such as those sold for cattle, is also an attractor.

Salamanders

A BACKYARD THAT OFFERS constantly moist areas, adequate undergrowth vegetation, and rock and wood piles – all in close proximity to one another – stands a strong chance of attracting one or more species of the common salamander. A shallow pool of water with vegetation growing in it also greatly enhances the appeal for these amphibians. Rainy, spring nights will often bring them into the open for easy viewing with the aid of a flashlight.

EASTERN NEWT

NOTOPHTHALMUS VIRIDESCENS

Although the greenish brown adult stage of the Eastern newt is strictly an aquatic creature, the immature stage – a bright red salamander with orange and black spots, known as the red eft – spends up to three years as a terrestrial form before returning to the water.

BELOW The red eft – the terrestrial, immature stage in the adult aquatic Eastern newt's development – has a bright orange coloration with distinctive red spots, encircled in black, along its back.

DATABOX

DESCRIPTION: Terrestrial eft is bright orange with red, black-encircled spots along sides and black dots on underside. 1¼ to 3½ inches long. Aquatic newt is dull brown-green with red, black-encircled spots along sides and black dots throughout. 2¼ to 5½ inches long.

HABITAT: Adults in lakes and ponds with heavy vegetation. Efts in woodlands near water.

REPRODUCTION/REARING: Breeds in late winter to spring. Female lays 100–400 individual eggs on vegetation in the water. Eggs hatch in 20–35 days. Larvae become efts in late summer and will remain in the terrestrial stage for 1–3 years. Some bypass the eft stage entirely.

HABITS: Efts are seen in large numbers on the forest floor after spring rains. Adults are common in shallow water.

HOW TO ATTRACT: Pond with vegetation for adults. Damp, wooded area with logs on ground for efts.

RED-BACKED SALAMANDER

PLETHODON CINEREUS

This very common species is the small salamander with the black-speckled red line down its back that is commonly discovered in basements during the wet months of early spring. The red-backed salamander is better able to survive dry conditions than most other species and thus is generally more common in developed areas.

DATABOX

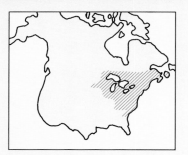

DESCRIPTION: Gray with broad red, black-bordered band along entire length of back; mottled black and white underside. Red band replaced at times with one of lead-gray to black. 2¼ to 5¼ inches long.

HABITAT: Moist woodlands.

REPRODUCTION/REARING: Mates fall to spring. Female lays cluster of 6–16 eggs under rock or log in mid-summer, then remains with the eggs until they hatch in 7–8 weeks. No aquatic stage. Larvae transform into adults in 2 years.

HABITS: Emerges at night to hunt. Remains underground during periods of little or no rain.

HOW TO ATTRACT: Moist areas with leaf litter and rotting logs.

SLIMY SALAMANDER

PLETHODON GLUTINOSUS

The appropriately named slimy salamander, a creature of wooded areas, secretes a sticky substance through its skin when threatened that is very difficult to remove, even with soap and water.

DATABOX

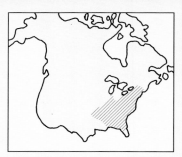

DESCRIPTION: Shiny black with yellow to cream-colored dots and spots scattered across body, more pronounced on underside. 4 to 7¼ inches long.

HABITAT: Moist wooded areas, shaded rocky areas, underground openings.

REPRODUCTION/REARING: Mates in spring in north, and summer in south. Female lays 6–30 eggs in burrow or under log in late spring in north and late summer in south, and then guards the nest until the larvae hatch in summer in the north and mid-fall in the south. No aquatic stage. Mature into adults in 3 years.

HABITS: Active dusk to dawn. Most active after rains.

HOW TO ATTRACT: Moist areas with leaf litter and logs on ground.

TIGER SALAMANDER

AMBYSTOMA TIGRINUM

One of the most common salamanders in the United States, the tiger salamander also is one of the first to be pushed from its home by man's development of the land. The large, tiger-striped amphibian insists on a moister environment with more ready access to a pool of water than many of the other common salamanders.

DATABOX

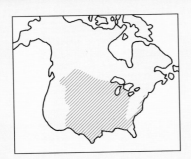

DESCRIPTION: Shiny brown-green with cream and black spots across entire body, or tiger-striped in dark on light or light on dark. 5¾ to 13½ inches long.

HABITAT: Extremely varied, but generally damp areas with soft ground.

REPRODUCTION/REARING: Mates and female lays more than 100 eggs from January to March in temporary and permanent sources of standing water. The eggs hatch in 3–4 weeks. The larvae transform to adult form in 75–120 days.

HABITS: Often seen on forest floor after heavy rains. Also found in leaf litter close to water.

HOW TO ATTRACT: Damp areas with plenty of leaf litter and soft soil.

SPOTTED SALAMANDER

AMBYSTOMA MACULATUM

Except for a few weekends each spring, when mating and egg-laying takes place in temporary woodland ponds, the spotted salamander spends its entire life (up to 11 years) in its underground tunnels.

DATABOX

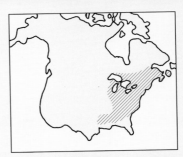

DESCRIPTION: Black to dark brown with two rows of orange dots along back; gray underside. 6 to 10 inches long.

HABITAT: Open forests near water.

REPRODUCTION/REARING: Mates December to early April (depending upon latitude). Female lays mass of about 100 eggs in breeding pond, where they attach to submerged vegetation. Larvae hatch in 5–8 weeks and transform into adults after another 8–15 weeks.

HABITS: Except for breeding, remains underground most of the year.

HOW TO ATTRACT: Moist, soft soil with leaf litter covering.

Pond Frogs

MOST POND FROGS, because of their moist skins, require a constant source of water in their immediate environment. A small pond, filled with growing vegetation and protected by overhanging vegetation, will attract some of these amphibians into many a backyard.

GREEN FROG

RANA CLAMITANS

Although the green frog is mostly a nocturnal species, it is generally not as quick to leap away from a perceived threat. The northern race of this species has a greenish skin, while the southern race has a tan skin.

DATABOX

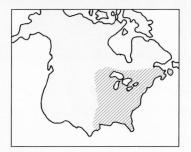

DESCRIPTION: Green to brown; white with gray lines and spots on underside. 2 to 4¼ inches long.
HABITAT: Ponds, lakes, swamps with submerged vegetation.
REPRODUCTION/REARING: Mates spring to mid-summer, when air temperature reaches 60°F plus. Female lays several clusters of eggs in water. The eggs hatch in 3–5 days. Tadpoles develop into adults in 370–400 days.
HABITS: Active dusk to dawn. Slow to leap in escaping danger. Voice is like a banjo note.
HOW TO ATTRACT: Aquatic habitat with submerged vegetation.

BULLFROG

RANA CATESBEIANA

With a possible length from tip of nose to tip of out-stretched leg of nearly 9 inches, the bullfrog is generally the largest frog to be encountered along North American waters. Originally restricted to the eastern half of the continent, the bullfrog's range has extended to many other areas through releases by man.

DATABOX

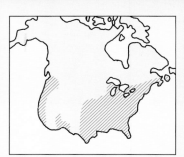

DESCRIPTION: Brown-green to yellow-green with mottling and scattered spots of brown or gray; white to buff underside. Large eardrum is visible behind eye.
HABITAT: Ponds, lake, streams, and rivers with underwater vegetation and mud.
REPRODUCTION/REARING: Mates late spring to mid-summer in north and late winter to mid-fall in south. Female lays egg masses, totalling as many as 20,000 eggs, that attach to underwater vegetation. Eggs hatch in 4 days. Tadpoles transform to adults in 1 year.
HABITS: Most active dusk to dawn. Croaks loudly.
HOW TO ATTRACT: Aquatic habitat with submerged vegetation mandatory.

WOOD FROG

RANA SYLVATICA

Each spring, large numbers of wood frogs descend on woodland ponds, all at about the same time, to breed, only to disappear back into the surrounding forest within a day or two. The wood frog is usually the first to start calling each spring, with its ducklike quack.

DATABOX

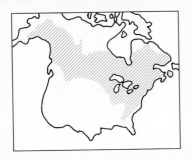

DESCRIPTION: Tan to dark brown with dark brown or black mask about eyes; white underside. 1¼ to 3¼ inches long.
HABITAT: Moist woodlands and grasslands.
REPRODUCTION/REARING: Mates late winter to early spring. Female lays egg masses in water. They attach to vegetation and hatch in 4–24 days.
HABITS: Active during the day. Voice is a ducklike quack.
HOW TO ATTRACT: Moist areas with dense vegetation.

PICKEREL FROG

RANA PALUSTRUS

Although it is regularly found much further from water sources than most of the other pond frogs, the pickerel frog must return to the pond each spring to breed and each fall to hibernate. Like most other frogs, the adult is carnivorous, eating any small creatures that it can catch.

DATABOX

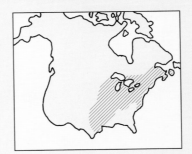

DESCRIPTION: Light tan with rows of green-brown to bronze patches and dots running front to back; white underside with yellow area near hindlegs.
HABITAT: Moist areas with dense vegetation near water.
REPRODUCTION/REARING: Mates late winter to spring. Female lays egg masses in water. The masses attach to vegetation and hatch in 6–20 days. Tadpoles develop into frogs in 1 year.
HABITS: Active dusk to dawn. Voice is long-lasting croak.
HOW TO ATTRACT: Correct habitat.

Treefrogs

MOST OF THE TREEFROGS that inhabit North America are quite small (less than 2 inches long). Most also are equipped with suction pads on their toes to enable them to climb about stems and twigs of various plants. Despite their family name, treefrogs generally need to live in close proximity to some constant water source.

SPRING PEEPER

HYLA CRUCIFER

The call of the spring peeper, which sounds like the clanging of a sharp bell, can carry as far as a mile on a warm spring night. Even the slightest pollution of its wetland home signals the end of the spring peeper in that location.

COMMON GRAY TREEFROG

HYLA VERSICOLOR

Spending most of its life high in the trees, the gray treefrog descends only at night and then only in the spring mating season. The trees that it calls home must be close to or actually growing in a water source that never completely dries up.

DATABOX

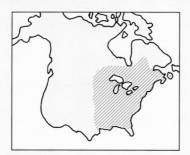

DESCRIPTION: Shades of brown to gray, with darker "X" across back; lighter underside. ¾ to 1½ inches long.
HABITAT: Woodlands in or near standing water.
REPRODUCTION/REARING: Mates and female lays 800–1,200 eggs late winter to early summer in north, late fall to late winter in south. Eggs hatch in 5–15 days. Tadpoles transform into adults in midsummer.
HABITS: Active dusk to dawn. Voice is like a small bell.
HOW TO ATTRACT: Correct habitat.

DATABOX

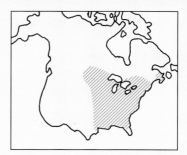

DESCRIPTION: Green, brown, or gray with irregular dark brown lines and patches; gray underside. 1 to 2½ inches long.
HABITAT: Larger vegetation in or near water.
REPRODUCTION/REARING: Mates and female lays eggs spring through summer. Eggs, as many as 2,000 per female, hatch in 4–6 days.
HABITS: Active dusk to dawn. Spends most of time in treetops. Voice is a trill.
HOW TO ATTRACT: Correct habitat.

CHORUS FROG

PSEUDACRIS TRISERIATA

At ponds where they congregate for breeding in very early spring, the constant calling sent up by the males can be nearly deafening. The chorus frog lacks the suction cups on its toes that most members of this frog family have.

DATABOX

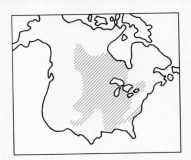

DESCRIPTION: Brown, green, or gray, with darker stripes along back and darker patches on legs; lighter underside. ¾ to 1½ inches long.
HABITAT: Grasslands and woodlands.
REPRODUCTION/REARING: Mates and female lays eggs late winter through summer in north, winter in south.
HABITS: Active dusk to dawn. Voice is a short trill.
HOW TO ATTRACT: Correct habitat.

NORTHERN CRICKET FROG

ACRIS CREPITANS

The cricket frog is generally found among the vegetation at the edge of some standing water source. When threatened it will quickly leap into the water, turn about while swimming and return to shore at a new location.

DATABOX

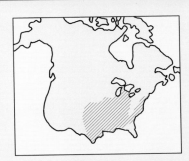

DESCRIPTION: Mottled tan and green, with spots of yellow, red and black, and dark brown bands across legs; white underside. ½ to 1½ inches long.
HABITAT: Waterways with thick vegetation but also exposure to the sun.
REPRODUCTION/REARING: Mates and female lays eggs spring through summer.
HABITS: Active during the day and highly elusive. Voice is like a click.
HOW TO ATTRACT: Aquatic habitat essential.

American Toad

BUFO AMERICANUS

NEARLY ANY HABITAT, from city backyard to mountain meadow, that offers loose soil and cool, damp shelter for the daytime will eventually attract an American toad or two.

The amphibian's appetite for harmful insect pests, combined with its harmless nature toward the garden plants, makes it a welcome resident in most backyard gardens. The toad will also come to low-standing lights at night to snatch insects that congregate in these places.

Contrary to popular folklore, holding a toad will not cause warts. However the amphibian does release a noxious liquid when grabbed that will irritate most smaller predators, like dogs and cats, causing them to avoid future encounters. The secretion does not seem to bother the hognose snake, which is a primary predator on the American toad.

At nearly the same time each spring, hordes of male toads will migrate from the surrounding woodland into area lakes and ponds, and begin their trilling calls that mark the beginning of the breeding season. They will be joined in a few days by the females, each one capable of laying as many as 15,000 eggs in a 70-foot string of jelly. All of the toads disperse within a few days of the egg-laying.

DATABOX

DESCRIPTION: Many different shades of brown with darker and light spots and covered with "warts;" tan underside is spotted with darker brown. 2 to 4½ inches long.

HABITAT: Extremely varied. Steady source of insect food and moist areas with moist, daytime-hiding cover are essential.

REPRODUCTION/REARING: Mates late winter to spring in water. Female lays strings of up to 15,000 eggs in water. Strings attach to submerged vegetation and eggs hatch in 3–14 days. Tadpoles transform into frogs in 40–60 days.

HABITS: Active dusk to dawn. Will visit low-standing lights for the insects they attract at night. Voice is a bird-like trill.

HOW TO ATTRACT: Provide moist, daytime-hiding cover and enhance the backyard environment for insects.

Painted Turtle

CHRYSEMYS PICTA

ONLY THOSE BACKYARDS that provide or are near relatively large ponds, lakes, or rivers with muddy bottoms will play host to the painted turtle.

This reptile most often will be seen sunning itself on exposed rocks and logs surrounded by water. When first discovered, the painted turtle will slip quietly into the water. But, within a few minutes, it will pop its head through the surface to check for danger and then climb slowly back onto the rock or log.

It spends the winter months in hibernation in the mud of the body of water, although sunny days that heat the water and mud may cause it to stir, even under the ice.

DATABOX

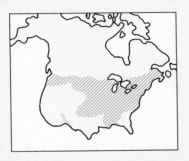

DESCRIPTION: Dark green upper shell with orange to yellow lines between segments and similarly colored bars or lines at edge; yellow lower shell; green head, neck, legs, and tail with yellow to red stripes. 4 to 10 inches long.
HABITAT: Standing or slow-moving water with muddy, vegetated bottoms and plenty of exposed rocks and logs.
REPRODUCTION/REARING: Clutch of 3 to 15 eggs in land-based nest cavity (but near water's edge) throughout summer. Two clutches at most each summer in north, as many as four in south. Eggs hatch within 3 months.
HABITS: Spends a great deal of daylight hours sunning itself on exposed rocks and logs.
HOW TO ATTRACT: Correct habitat.

Box Turtle

EASTERN – TERRAPENE CAROLINA; WESTERN – TERRAPENE ORNATA

THE SLOW-MOVING BOX TURTLE is the species most often captured and kept as pets by small children. It is also the principal reptilian target of the senseless practice of carving names and dates into the shell.

Whenever threatened, the box turtle's response is to pull its head, legs, and tail inside its shell and close up the openings behind it. It is able to close more completely than most turtle species because its lower shell is hinged to allow the front and rear sections to bend to meet the edges of the upper shell.

It is an extremely long-lived creature. Some have been reported to be more than 100 years old, although few systems of aging are completely reliable.

DATABOX

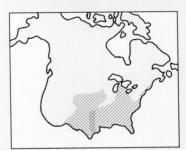

WESTERN

EASTERN

DESCRIPTION: Brown-green upper shell with yellow to orange patches, dots, and lines radiating across each segment. Yellow, orange, brown, or green lower shell is jointed to allow complete closure with upper shell. Head, neck, legs, and tail are brown to green. 4 to 8 inches long.
HABITAT: Moist woodlands, grasslands and agricultural areas.
REPRODUCTION/REARING: One clutch of 3–10 eggs laid in land-based nest May through July. The eggs hatch in 3 months. One mating enables female to lay fertile eggs for several subsequent summers.
HABITS: Terrestrial. Limited to small home range.
HOW TO ATTRACT: Correct habitat. Relocation may cause turtle to adopt new home range, if habitat and food needs are met.

Green Anole

ANOLIS CAROLINENSIS

ALTHOUGH THE ANOLES are the largest family of reptiles in the Americas, with some 200 different species, only the green anole is native to North America. A few of the other species have been introduced by man into Florida, but with limited success.

Like the more famous chameleon, anoles can change color in response to environmental conditions as well as states of mind. At rest they are generally brown. But males battling over territory, or either sex when frightened or excited, become green.

They eat a wide variety of small insects, which they stalk and pounce upon in catlike fashion. They are extremely agile and swift climbers.

DATABOX

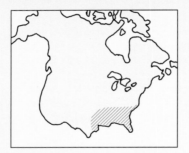

DESCRIPTION: Shades of green with small cream to gray-blue spots, and a rose-colored throat fan, but can change to shades of brown as well; lighter underside. 4¾ to 8½ inches long.
HABITAT: Off-ground levels of trees, shrubs, weeds and grasses.
REPRODUCTION/REARING: Mates late winter through late summer with individual eggs laid in debris or rock piles thereafter at 2-week intervals. Eggs hatch in 1–2 months.
HABITS: Active during the day. Spends a lot of the daylight hours basking in the sun.
HOW TO ATTRACT: Correct habitat, including rock piles or walls.

Eastern Fence Lizard

SCELOPORUS UNDULATUS

THE EASTERN FENCE LIZARD provides a ready study in adaptation to the environment.

In the eastern part of the continent, with its many forested areas, the lizard is largely a tree dweller, quick to escape in an upward direction. But, as we move west to the grassland and prairie regions, the small lizard becomes predominantly a ground dweller, seeking its hiding places under brush and underground.

The lizard's diet is likewise variable. It will eat almost any smaller creature.

DATABOX

WESTERN

EASTERN

DESCRIPTION: Shades of brown, gray, and green with indistinct, darker "V"s along back; lighter underside with blue mottling. Eastern fence lizard is 3½ to 7 inches long; western, 6 to 9 inches long.
HABITAT: Eastern – open wooded or grassy areas that offer sun exposure. Western – wooded or rocky areas that offer sun exposure.
REPRODUCTION/REARING: Eastern – mates spring through summer. Female lays 1–3 clutches of 3–12 eggs each. Eggs hatch mid- to late summer. Western – mates in early spring. Female lays 1 clutch of 3–12 eggs, which hatch in late summer.
HABITS: Active during the day. Often seen on fence posts and rock walls.
HOW TO ATTRACT: Correct habitat.

Five-lined Skink

EUMECES FASCIATUS

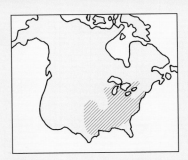

DESCRIPTION: Brown to black with five lighter stripes running head to tail; lighter underside. 5 to 8 inches long.

HABITAT: Damp, wooded areas with plenty of rotting ground debris.

REPRODUCTION/REARING: Mates April to May. Female lays clutch of 5–14 eggs in small burrow and tends them until they hatch in June to July.

HABITS: Active during the day. Spends much of its time on ground, but climbs tree trunks and rocks to sun itself.

HOW TO ATTRACT: Correct habitat.

ALTHOUGH THE FIVE-LINED SKINK is a lizard (a reptile), its preference for cool, moist spots such as in leaf litter and under old logs aligns it with the salamanders (amphibians). However, it retains the lizard's liking of open areas for sunning, so long as those spots are near the leaf litter and old logs.

Immature individuals are colored so differently from adults that they were once thought to be a separate species. Notable among the juvenile coloration is a bright blue tail.

The five-lined skink is active during the daylight hours, as it hunts for insects and other small creatures.

BELOW In the adult five-lined skink, the black/brown coloration is divided by five lighter colored stripes running the entire length of its body – hence its name.

Ringneck Snake

DIADOPHIS PUNCTATUS

THE SMALL RINGNECK SNAKE is one of the more common snakes across much of the United States and often frequents backyards, but its highly secretive nature generally allows it to pass undetected. Most often the snake is found when a weedy area is being tidied up.

When first handled it will strike, but it is non-poisonous and its jaws are too small and weak to inflict injury on a human. After a few attempts it will generally calm down.

DATABOX

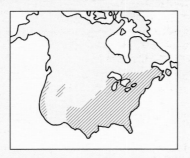

DESCRIPTION: Gray to black with a single red to orange band around neck; yellow to red underside with regular rows of black dots. 9 to 30 inches long.
HABITAT: Moist wooded and grassy areas.
REPRODUCTION/REARING: Mates spring or fall. Several females lay their clutches of 2–9 eggs together. Eggs hatch within 2 months and young reach maturity within 3 years.
HABITS: Active day and night. Very elusive.
HOW TO ATTRACT: Correct habitat.

Rat Snake

ELAPHE OBSOLETA

AN EFFICIENT PREDATOR ON RATS, mice, and similar rodents, the rat snake can be a welcomed guest in man's backyard. However, it is an excellent climber and includes a great many birds and bird eggs in its diet.

The rat snake will be active in the daytime in spring and fall, but takes on nocturnal habits to avoid the heat of the summer day.

DATABOX

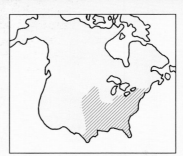

DESCRIPTION: Can appear in one of three colorations: black with off-white between scales; gray to brown with darker patches; and brown, red, or yellow with darker stripes, all with lighter underside. 3 to 8½ feet long.
HABITAT: Woodlands, grasslands, and agricultural areas.
REPRODUCTION/REARING: Mates in spring. Female lays 6–32 eggs under rocks or logs in summer. Eggs hatch in 2–4 months.
HABITS: Active during the day in cooler months, dusk to dawn in mid- to late summer. An able climber.
HOW TO ATTRACT: Not readily attracted.

Hognose Snake

EASTERN – HETERODON PLATYRHINOS
WESTERN – HETERODON NASICUS

WHEN THREATENED, THE HOGNOSE SNAKE puffs out its neck in a cobra-like fashion, hisses, and strikes at the attacker. It rarely bites humans, however. If its false bravado fails, the snake will play dead, rolling onto its back, holding its mouth open with the tongue hanging out and making its entire body limp and lifeless.

Apparently immune to the noxious secretions of many toad species, the hognose snake is a principal predator on the amphibians. It also eats a wide range of other small creatures.

DATABOX

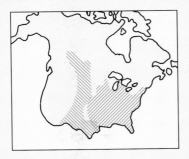

WESTERN

EASTERN

DESCRIPTION: Shades of brown, gray, yellow, or red, with large patches or bands of darker shades; underside is lighter and mottled. 18 to 46 inches long.
HABITAT: Open areas with loose, sandy soil.
REPRODUCTION/REARING: Mates spring or fall. Female lays 6–60 eggs in a burrow in summer. Eggs hatch in 5–6 weeks.
HABITS: Active during the day. When threatened, puffs out neck and hisses. If that fails, it rolls onto its back and plays dead.
HOW TO ATTRACT: Not readily attracted.

Milk Snake

LAMPROPELTIS TRIANGULUM

FOLKLORE HOLDS THAT THE MILK SNAKE will drink milk directly from the udders of cows, but the reptile's actual diet is closer to that of other snake species: rodents, birds, lizards, frogs, and other snakes (including poisonous species like the rattlesnake). It kills by grasping the victim behind the head, coiling about it and constricting until the prey suffocates.

The non-poisonous milk snake gains a measure of protection from species that would prey upon it by mimicking the coloration of poisonous species: the copperhead in the east and the coral snake in the south.

DATABOX

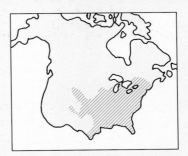

DESCRIPTION: Ringed or spotted in combinations of gray, tan, black, yellow, red, or orange. 15 to 78 inches long.
HABITAT: Extremely varied.
REPRODUCTION/REARING: Mates in spring. Female lays 2–20 eggs in or under rotting log in early summer. Eggs hatch in 1½ to 2 months.
HABITS: Active dusk to dawn. Highly elusive.
HOW TO ATTRACT: Not readily attracted.

Northern Water Snake

NERODIA SIPEDON

RARELY VENTURING FAR from the water's edge, the water snake is easily aggitated and inflicts a painful bite that can lead to infection. It is most often observed coiled in the branches of nearly submerged trees and shrubs on warm, sunny days.

The water snake is a powerful swimmer and includes fish and frogs as staples in its diet.

DATABOX

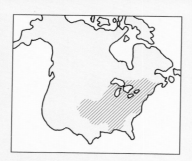

DESCRIPTION: Shades of brown or gray with darker patches along back and sides; off-white underside. 22 to 56 inches long.
HABITAT: Extremely varied, aquatic.
REPRODUCTION/REARING: Mates in late spring. Female gives birth to 15–50 young in late summer.
HABITS: Active any time. Often seen sunning itself on exposed rocks or logs, or crawling through vegetation at shoreline.
HOW TO ATTRACT: Not readily attracted.

Common Garter Snake

THAMNOPHIS SIRTALIS

THE COMMON GARTER SNAKE is the most commonly encountered of all snake species in North America. It is also quick to strike any aggressor, including man, as well as excreting a foul-tasting liquid.

Although most specimens are much smaller, the common garter snake will grow to a length of 3 feet. The female bears her 10–60 young alive.

DATABOX

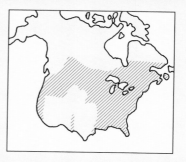

DESCRIPTION: Coloring is highly variable, but generally shades of green with distinct, lighter stripes along sides and back; lighter underside. 16 to 53 inches long.
HABITAT: Moist areas near water.
REPRODUCTION/REARING: Mates in late winter to spring. Female gives birth to 6–80 young in late summer to fall. The young mature within 2 years.
HABITS: Active during the day. Quick to strike when threatened.
HOW TO ATTRACT: Not readily attracted.

Smooth Green Snake

OPHEODRYS VERNALIS

THE SMOOTH GREEN SNAKE is a very docile species, extremely reluctant to strike when handled. Its diet consists mostly of grasshoppers, spiders, and caterpillars.

The female lays her eggs under sun-warmed stones, where they hatch in 4–25 days, an extremely short time-span for the hatching of snake eggs.

BELOW The vivid coloring of the smooth green snake is unmistakable. This snake was photographed in its preferred – moist and weedy – habitat.

DATABOX

DESCRIPTION: Bright green; white underside; very smooth scales. 14 to 28 inches long.
HABITAT: Moist, grassy, weedy areas.
REPRODUCTION/REARING: Mates in spring to summer. Female lays 2–10 eggs, often with other females, in late summer. Young hatch in 1–3 weeks.
HABITS: Active in the day. An able climber.
HOW TO ATTRACT: Not readily attracted.

Bald-faced Hornet

VESPULA MACULATA

ONE OF THE MOST COMMON and widespread wasps in North America, the bald-faced hornet's presence can be easily detected by the paper-like nest that it constructs under protected overhangs and tree limbs.

Although each nest is filled with thousands of larvae, which the workers keep nourished with many types of insect prey, only the queens survive from year to year. The queens disperse from the nest each fall, hibernate individually over the winter and then begin new nests when they emerge the following spring.

DATABOX

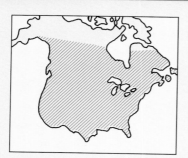

DESCRIPTION: Black with cream to white patterns across head and at rear of abdomen; wings are smoked black. ½ to 1 inch long.

SIGN: Rounded nest with a paper-like appearance, hanging in sheltered location.

HABITAT: Open grassy, weedy areas.

REPRODUCTION/REARING: Mated female begins nest in spring, producing female workers who supply the larvae with insects. Males develop late in summer to mate. Only females that have mated survive through the winter.

HABITS: Extremely defensive of nest.

HOW TO ATTRACT: Because of the insect's defensive and stinging nature, it is not a species to be attracted intentionally. But it will visit flowers and hummingbird feeders.

Yellow Jacket

EASTERN – VESPULA MACULIFRONS
WESTERN – VESPULA PENNSYLVANICA

COMMONLY SEEN AT PICNICS, where they can become persistent pests as they attempt to snatch and carry off tiny bits of food, yellow jackets can deliver painful stings. Unlike many stinging insects, female yellow jackets can sting repeatedly.

They generally nest in the ground or hollow logs and stumps, and are aggressive defenders of the nest site.

DATABOX

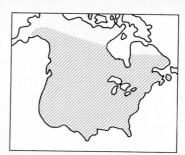

DESCRIPTION: Black with yellow to white bands; wings are smoked black. ½ to ¾ inches long.

SIGN: Nest is usually in ground, log, or stump.

HABITAT: Grassy, weedy areas.

REPRODUCTION/REARING: Similar to bald-faced hornet.

HABITS: Appear at nearly every outdoor event involving food. Quick to sting.

HOW TO ATTRACT: Not a species to be attracted, but it will visit flowers and hummingbird feeders.

American Bumble Bee

BOMBUS PENNSYLVANICUS

THIS IS THE LARGE, furry black and yellow bee commonly seen buzzing slowly among many species of flowers in the garden. The peaceful appearance of individual bees belies the readiness of the colony to attack and pursue any trespasser that ventures to near their ground-based nest.

These bees will use nest boxes that are provided with ½-inch entrance holes and lined on the inside with cotton.

DATABOX

DESCRIPTION: Black with yellow bands; broad, hairy body; wings are smoked black. ¼ to ¾ inches long.
SIGN: Nests underground.
HABITAT: Open areas.
REPRODUCTION/REARING: Queen begins nest in spring, first producing workers and later males and potential queens. Only mated females survive through the winter.
HABITS: Slow-flying. Powerful sting.
HOW TO ATTRACT: Flowers.

Honey Bee

APIS MELLIFERA

ALTHOUGH THE HONEY BEE is now spread across all but the northernmost reaches of North America, it is not a native species. It was brought to the New World by pioneer settlers in the 17th century.

This is the species of bee for which bee-keepers construct the small, multi-sectioned box-hives that can be seen at field edges across the continent.

DATABOX

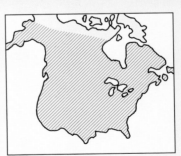

DESCRIPTION: Brown to black with yellow to orange bands; wings are translucent. ¼ to ¾ inches long.
SIGN: Builds hives in hollow trees.
HABITAT: Open grassy and wooded areas.
REPRODUCTION/REARING: Queen lays eggs that generate a colony of as many as 75,000 workers, which keep the queen fed on royal jelly. Larvae are fed a mix of pollen and honey. New queens emerge late spring or early summer when the old queen leaves with hundreds or thousands of workers to begin new colony. First new queen kills other queens, mates after a week or two and returns to lay eggs for new colony.
HABITS: Often seen in swarms on trees.
HOW TO ATTRACT: Flowers.

Orange Sulphur

COLIAS EURYTHEME

THE ORANGE SULPHUR continues to colonize new territory within its already coast-to-coast range, because of its ability to use many different legume species as host plants and nectar sources.

Also known as the alfalfa butterfly, this species produces two broods each summer.

DATABOX

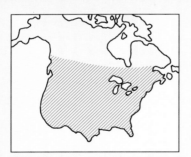

DESCRIPTION: Adult's wings are orange with a hint of pink and a black edge. Wingspan is 1½ to 2½ inches.
SIGN: Dark green, pink-striped caterpillar and green, yellow-dotted chrysalis are found on clovers and alfalfa.
HABITAT: Open areas.
FLIGHT OF ADULT: Late winter through late fall.
HABITS: Strong flier.
HOW TO ATTRACT: Clovers and alfalfa.

Cabbage White

PIERIS RAPAE

A HIGHLY DESTRUCTIVE SPECIES (in its green caterpillar stage) on garden and crop crucifers, such as cabbage and turnip, the cabbage white was introduced into Quebec about 1860. It quickly spread across North America and is among the most common residents of any area where crops are grown.

DATABOX

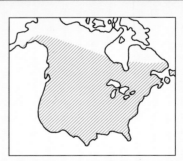

DESCRIPTION: Adult has white wings with black tips. Wingspan is 1 to 2 inches.
SIGN: Green, yellow-striped caterpillar and green, dotted with tan chrysalis are found on species of crucifers, such as cabbage.
HABITAT: Extremely varied.
FLIGHT OF ADULT: Early spring through late fall.
HABITS: Often seen flying in small groups.
HOW TO ATTRACT: A pest; not to be attracted.

Monarch

DANAUS PLEXIPPUS

THE MONARCH IS THE ONLY North American butterfly that engages in true bird-like spring and fall migration. Some populations cover 2,000 miles on the trip, migrating between Canada and Mexico. The butterflies that return in the spring are not the same individuals that left in the fall, but rather new generations that were produced during the northward migration.

Few predators prey upon the monarch because of the noxious taste it aquires while a caterpillar feeding on milkweed.

DATABOX

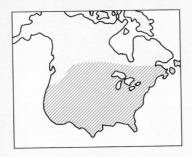

DESCRIPTION: Adult has orange wings with strong veins of black. Wingspan is 3 to 4 inches.
SIGN: Black, white, and yellow-banded caterpillar and green, gold-dotted chrysalis are found on milkweed plants.
HABITAT: Near milkweed.
FLIGHT OF ADULT: Early summer to fall.
HABITS: Migratory.
HOW TO ATTRACT: Milkweed.

Mourning Cloak

NYMPHALIS ANTIOPA

UNLESS STARTLED FROM its resting place on the side of a tree, the mourning cloak's camouflage is virtually perfect. When it does take flight, it reveals a sharply contrasting underside of yellow bands and blue spots that often startles a would-be predator and allows the butterfly to escape.

Rotting fruit, carrion, and scat are like magnets to any mourning cloaks in a given area.

DATABOX

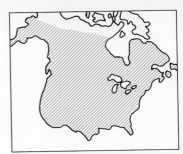

DESCRIPTION: Adult has dark brown wings with a band of shiny blue patches and cream-colored edges. Wingspan is 2¾ to 3½ inches.
SIGN: Black, red-spotted caterpillar and gray to brown chrysalis are found on many deciduous trees, including cottonwood, willow, and elm.
HABITAT: Open, wooded areas.
FLIGHT OF ADULT: Late winter through late fall.
HABITS: Often found resting on side of tree.
HOW TO ATTRACT: Rotting fruit, tree sap, scat, and carrion.

American Copper

LYCAENA PHLAEAS

ABUNDANT IN ALL but the far south and west United States, the American copper is aggressively defensive of its home territory. Males will attack much larger butterflies, birds, dogs, and even humans in attempts to drive the trespasser off.

Host plants for the caterpillar of this species include several sorrels and docks.

Tiger Swallowtail

PAPILIO GLAUCUS

OFTEN SEEN IN GROUPS at mud puddles, the tiger swallowtail is also attracted to carrion and smoke. It is one of the most common butterflies throughout its range.

The green caterpillar sports two huge, false eyespots on its head. When these spots fail to thwart a predator and the caterpillar is grabbed, it shoots two orange horns from its head, which will retract when it is no longer threatened.

DATABOX

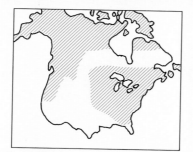

DESCRIPTION: Adult has orange wings with brown to black border and black patches. Wingspan is ¾ to 1¼ inches.
SIGN: Green, red- or yellow-marked caterpillar and green, chrysalis are found on sorrel.
HABITAT: Grassy, weedy areas.
FLIGHT OF ADULT: Spring to mid-summer.
HABITS: Slow-flying.
HOW TO ATTRACT: Sorrel.

DATABOX

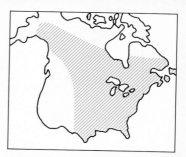

DESCRIPTION: Adult has yellow wings with black tiger stripes. Wingspan is 4 to 6 inches.
SIGN: Green caterpillar (with yellow eyespots, and black and yellow bands) and green chrysalis are found on ash, poplar, willow, and birch.
HABITAT: Open, wooded areas.
FLIGHT OF ADULT: Spring through late summer.
HABITS: Often seen in groups at mud puddles.
HOW TO ATTRACT: Many types of flowers.

Silver-spotted Skipper

EPARGYREUS CLARUS

THE SILVER-SPOTTED SKIPPER is one of the most widespread of all butterflies on the North American continent. It can be found in almost any environment, from city garden to open meadow.

The caterpillar spends the daylight hours in a nest made of a curled leaf, emerging to feed at night.

DATABOX

DESCRIPTION: Adult has deep brown wings with yellow patches and spots, and a white edge. Wingspan is 1¾ to 2¼ inches.

SIGN: Orange to brown-red caterpillar is found on beans, locusts, and wisteria. The brown chrysalis is buried in leaf litter.

HABITAT: Open areas.

FLIGHT OF ADULT: Summer.

HABITS: Often seen flying acrobatically, high in the air.

HOW TO ATTRACT: Many types of flowers.

BELOW A silver-spotted skipper is captured on film feeding on swamp milkweed (*Asclepias incarnata*).

Isabella Tiger Moth

ISIA ISABELLA

BETTER KNOWN FOR ITS CATERPILLAR, the banded woollybear, this moth is an extremely widespread species. The caterpillar is held up as a weather forecaster by popular folklore, allegedly showing the harshness of the coming winter by its amount of black banding.

When threatened, the caterpillar plays dead by curling into a tight coil and remaining still. It hibernates at the caterpillar stage.

DATABOX

DESCRIPTION: Adult has brown to yellow-brown wings with black dots. Wingspan is 1½ to 2 inches.
SIGN: Red-brown and black-banded, woollybear caterpillar seen in open space.
HABITAT: Weedy, grassy areas.
FLIGHT OF ADULT: Summer in north, late winter through fall in south.
HABITS: Feeds on herbaceous plants.
HOW TO ATTRACT: Not easily attracted.

Clear-winged Sphinx Moth

HEMARIS THYSBE

THE CLEAR-WINGED SPHINX MOTH is also known as the hummingbird moth because of its habit of hovering over flowers like a hummingbird to take nectar. Its wings produce a buzzing sound similar to, but softer than, that of the hummingbird. The white-lined sphinx (*Hyles lineata*) shares this characteristic.

DATABOX

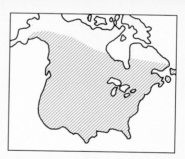

DESCRIPTION: Adults have mostly clear wings, except along veins, where the scales have not dropped off. Wingspan is 1½ to 2¼ inches.
SIGN: Yellow-green caterpillar (with darker lines and spot) is found on honeysuckle.
HABITAT: Grassy and weedy areas.
FLIGHT OF ADULT: Spring through summer.
HABITS: Hovers over flowers like a hummingbird.
HOW TO ATTRACT: Many types of flowers.

Cecropia Moth

HYALOPHORA CECROPIA

THIS LARGEST NORTH AMERICAN MOTH can sport a wingspan of 5½ inches. The caterpillar can grow to 4½ inches.

It is also a common species from the Rocky Mountains to the East Coast, wherever some of the 50 different tree species that the caterpillar uses as host plants occur. Those tree species include some of the continent's most common, such as birch, maple, ash, willow, elm, apple, lilac, and cherry.

DATABOX

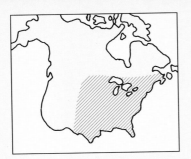

DESCRIPTION: Adult has red-brown wings with white and black spots, and red bands. Wingspan is 5 to 6 inches.

SIGN: Large green caterpillar, with blue tint on sides and many yellow tubercles. Can be found on willow, apple, elm, ash, maple, and lilac.

HABITAT: Open areas.

FLIGHT OF ADULT: Late spring or early summer.

HABITS: Cocoon attached along one side to a twig.

HOW TO ATTRACT: Host plants for caterpillars.

Luna Moth

ACTIAS LUNA

THIS LARGE, DELICATELY COLORED MOTH has long been a favorite of amateur collectors. Commonly found around street lights at night, it is easily captured. However, the luna moth's numbers have tumbled drastically in recent years, thanks to pesticides and environmental destruction.

DATABOX

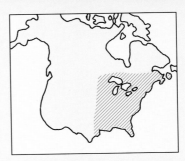

DESCRIPTION: Adult has pale green wings with purplish band along front and one vivid eyespot on each wing. Wingspan is 3 to 4½ inches.

SIGN: Adult flying around street lights at night.

HABITAT: Deciduous woodlands.

FLIGHT OF ADULT: Spring to early summer.

HABITS: Common species at street lights.

HOW TO ATTRACT: Host plants for caterpillar include birch, walnut, persimmon, and hickory.

Pennsylvania Firefly

PHOTURIS PENNSYLVANICUS

MANY A CHILDHOOD'S summer evening has been spent filling a glass jar with these small, soft-bodied beetles to create a natural lantern of sorts. The flashing tail lights (every three or four seconds while in flight) are the result of bioluminescence, which creates light without heat. Part of the insect's courtship ritual, the light is generated by an enzyme that glows when exposed to oxygen.

Although the firefly still exists in healthy numbers, the heavily pesticided and tightly mown lawns of today's suburbia produce many fewer than were present for previous generations of jar-toting youngsters.

DATABOX

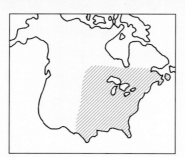

DESCRIPTION: Head yellowish with black spot at center and a pink edge; hard forewings are brown with yellow edges; yellow to green luminous organ at tip of abdomen. ¼ to ¾ inches long.

SIGN: Flashing in the air at night in summer.

HABITAT: Weedy and grassy areas.

REPRODUCTION/REARING: Larvae hatch in spring. Adults emerge in summer. Eggs laid in moist debris in late summer.

HABITS: All life stages are luminous.

HOW TO ATTRACT: Do not use pesticide on the lawn and do not mow the lawn to a very short length.

Dragonflies & Damselflies

ORDER – ODONATA

BETWEEN 400 AND 500 SPECIES of dragonflies and damselflies occur across North America. Nearly every locale on the continent has at least a few species. But, the dedicated enthusiast aside, most of us know them all simply as dragonflies.

Highly effective predators, they capture their prey in a hanging basket formed by their legs while in flight. Many pest species, such as gnats and mosquitoes, are primary prey for these aerial raiders. In the south certain species have been known to have devastating impact on bee-hives, earning them the local name of bee hawks.

They generally make their homes around open ponds and lakes that have much emergent vegetation, but they also commonly frequent backyards in search of prey.

DATABOX

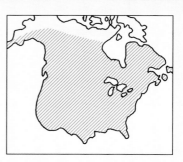

DESCRIPTION: Different species show immense color variation, but most have a spindly body, large compound eyes, and four elongated, transparent, heavily-veined wings. Wingspans vary from 1¾ to 5 inches.

HABITAT: Varied, but usually near water.

REPRODUCTION/REARING: Mates in flight, male and female coupled by male's clasper around female's neck. Female deposits eggs into water.

HABITS: Darts from one perch to another at water's edge.

HOW TO ATTRACT: Correct habitat.

Striped Blister Beetle

EPICAUTA VITTATA

THE STRIPED BLISTER BEETLE is among the most common of this family (about 200 North American species) of small beetles, which acquired their family name from the blister-causing chemical cantharidin that they carry in their bodies and can secrete when threatened.

Although the adult beetles can be major agricultural pests, the larvae feed on grasshopper eggs, each one destroying as many as three dozen of the eggs by the time they pupate.

DATABOX

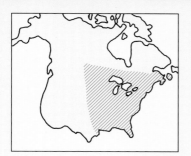

DESCRIPTION: Striped orange and dark brown or black front to back. Bulging head. ½ to ¾ inches long.
HABITAT: Agricultural areas.
REPRODUCTION/REARING: Adults emerge in late spring. Females lay clusters of as many as 100 eggs in the soil near grasshopper eggs. Eggs hatch within 3 weeks and larvae burrow into grasshopper eggs. The larvae pupate within 2 weeks and spend the winter months in the ground.
HABITS: Secretes an irritating liquid.
HOW TO ATTRACT: Not generally a species to be attracted.

Squash Bug

ANASA TRISTIS

A MAJOR AGRICULTURAL and garden pest, the squash bug is a common resident of nearly all of North America. It can be found in large numbers wherever plants of the squash family are grown. The insect damages the plants by boring into the leaves and stems and sucking out their vital juices.

Several species of tachinid flies (Family Tachinidae) are effective parasites on both the nymph and adult stages, and can help to reduce the squash bug's numbers. Man can also affect the local population by removing all plant litter from the garden after the harvest, thus eliminating the shelter that the adults need to hibernate through the winter.

DATABOX

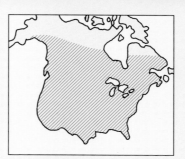

DESCRIPTION: Gray to brown, with darker edges along abdomen; lighter underside. ½ to ¾ inches long.
HABITAT: Agricultural areas and gardens.
REPRODUCTION/REARING: Females lay eggs in clusters of 15–45 on stems of host plants of the squash family. Nymphs feed on the plant juices.
HABITS: Adults seek shelter in large numbers in plant debris and leaf litter.
HOW TO ATTRACT: Not a desirable species.

Small Eastern Milkweed Bug

LYGAEUS KALMII

LIKE OTHER INSECT SPECIES that feed on the milkweed plant – notably the caterpillar of the monarch butterfly – milkweed bugs acquire a toxic or noxious taste from the plant that affords them protection from would-be predators.

DATABOX

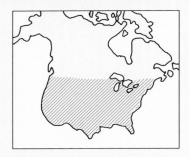

DESCRIPTION: Black with red dot on head and red "X" on forewings. ¼ to 1½ inches long.
HABITAT: Grassy, weedy areas; always near milkweed.
REPRODUCTION/REARING: As leaves and buds begin to appear on the milkweed, the female lays her eggs on the plants. Nymphs feed on the milkweed.
HABITS: Always found near milkweed.
HOW TO ATTRACT: Milkweed.

Colorado Potato Beetle

LEPTINOTARSA DECIMLINEATA

MORE COMMONLY KNOWN as the potato bug, this beetle is found across North America. The fat, pink caterpillars are highly destructive to the leaves of their namesake plant, but also other cultivated plants like tomato, eggplant, pepper, and tobacco. Before man provided these plants, wild species of the nightshade family were the principal host plants.

DATABOX

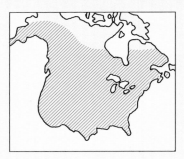

DESCRIPTION: Head is orange with black spots; hard forewings are striped cream to yellow and black. ¼ to ½ inch long.
HABITAT: Agricultural areas.
REPRODUCTION/REARING: Female lays cluster of eggs on underside of plant leaves in spring. Larvae pupate within 3 weeks and adults emerge in another 2 weeks.
HOW TO ATTRACT: A destructive pest, not to be attracted.

Thin-legged Wolf Spider

GENUS PARDOSA (MORE THAN 100 SPECIES)

LIKE OTHER MEMBERS of this family of spiders (*Lycosidae*), webs and other constructions commonly associated with spiders are foreign to the thin-legged wolf spider. True to its name it hunts across a small home territory, using both stalking and ambush techniques. Much of the hunting is done by night, although wolf spiders can be active during the daylight hours and spend a great deal of time basking in open, sunny spots.

Females attach their egg sacs to the spinnerets at the rear of their body and carry them with them until the young spiders hatch. The offspring then are carried on the female's back until they have reached the point at which they can fend for themselves.

The eight eyes of the wolf spider, arranged in three rows, makes it easy to locate with a flashlight. With magnification it has an other-worldly appearance.

DATABOX

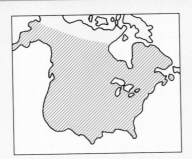

DESCRIPTION: Solid gray, hairy on body; dark and light stripes around legs; two rows of four eyes each. ¼ to ½ inch long.
HABITAT: Grassy areas.
REPRODUCTION/REARING: Female carries eggs in cocoon behind her, or once tiny spiderlings, on her back.
HABITS: Active during the day, as roams grassy territory hunting for prey.
HOW TO ATTRACT: No pesticides.

Grass Spider

GENUS AGELENOPSIS

GRASS SPIDERS construct their webs from grass stem to grass stem, one sheet of non-adhesive silk above another, which has a funnel off to one side. Their webs can easily be seen on any dew-coated morning, beginning in May.

When an insect flies into the top sheet, it is knocked down onto the lower sheet and the spider darts from its hiding place in the funnel to inject its prey with a venomous bite and then drag it back into the funnel for eating.

The grass spider lives for only one year, dying soon after the eggs are laid in the fall.

DATABOX

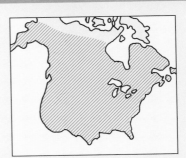

DESCRIPTION: Red-brown with alternating light and dark stripes running front to back; hairy legs. ½ to ¾ inch long.
SIGN: Small, funnel-shaped webs in the grass.
HABITAT: Grassy areas.
REPRODUCTION/REARING: Eggs pass the winter months in an egg sac among debris. Spiderlings hatch and disperse in spring.
HABITS: A quick-running spider.
HOW TO ATTRACT: No pesticides.

Garden Spider

ARANEUS DIADEMATUS

THE GARDEN SPIDER, a species introduced from Europe in the 18th century, spins a new web every night, after eating the remains of the one it used that day to capture its prey.

It is part of the family of orb-weaving spiders (*Araneidae*), which spin the webs of concentric lines of silk that we most commonly identify with as spider webs. The strands of the web have an adhesive quality that holds any insect that comes into contact with them in place until the spider can inject it with venom and wrap it in a silken cocoon for storage and later consumption.

BELOW A garden spider approaches its prey, captured in the sticky silk threads of the orb web that the spider spins afresh each night.

DATABOX

DESCRIPTION: Brown with tiny silver-gray dots; orangish legs. ¼ to ¾ inch long.

SIGN: Large, orb-type webs.

HABITAT: Open grassy and weedy areas.

REPRODUCTION/REARING: Female lays egg mass attached to twig at one side of her web.

HABITS: Eats old web and rebuilds every night.

HOW TO ATTRACT: No pesticides.

Field Cricket

GRYLLUS PENNSYLVANICUS

THIS WIDESPREAD INSECT and the nearly as widespread house cricket (*Acheta domestica*) are the species that we most often find sharing our homes with us. The field cricket generally waits until fall to move in, but once there will set up a chirping din equal to that of its cousin. Both can also cause damage to woolen and fur garments.

Being a cold-blooded creature, the cricket responds quickly to changes in temperature and reflects its responds in its chirping. The call grows faster as the temperature rises.

DATABOX

DESCRIPTION: Dark brown to black; long antennae. ½ to 1 inch long.
SIGN: Voice is a repeated series of three chirps.
HABITAT: Moist grassy and weedy areas.
REPRODUCTION/REARING: Female lays eggs into soil, where they spend the winter.
HABITS: Constant evening chirping.
HOW TO ATTRACT: No pesticides.

True Katydid

PTEROPHYLLA CAMELLIFOLIA

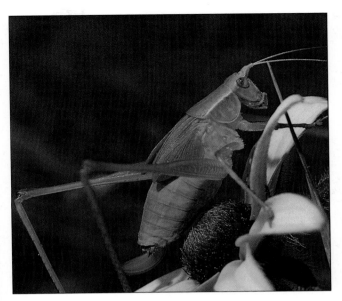

ALSO KNOWN AS THE NORTHERN KATYDID, this medium-sized green insect was named for its call, which clearly states, "katy-did." In some instances, the call might also be "katy-didn't." Both sexes call and the chorus reaches its peak at the height of the mating season in late summer.

Trees and shrubs are the typical calling location, but it is not all uncommon for a katydid to alight on a window screen at night and sing out from there.

DATABOX

DESCRIPTION: Bright green with distinct veins throughout forewings. 1¾ to 2¼ inches long.
SIGN: Voice is a distinct "Katy-did."
HABITAT: Wooded areas.
REPRODUCTION/REARING: The female lays eggs on the bark of deciduous trees in the fall. The eggs overwinter and hatch in late spring.
HABITS: Often land on window screen at night.
HOW TO ATTRACT: Must have wooded area, where no pesticide is used.

American Bird Grasshopper

SCHISTOCERCA AMERICANA

A CLOSE RELATIVE of the desert locust that occasionally devastates vast areas of crop and forest land in Africa and Asia, the American bird grasshopper is so named because of its strong flying ability.

This insect is unable to withstand extremes of temperature. It migrates north in large numbers early each summer and back south each fall.

The American bird grasshopper is an extremely wary insect, buzzing into flight at the least provocation and flying a healthy distance before disappearing into a tree or shrub.

DATABOX

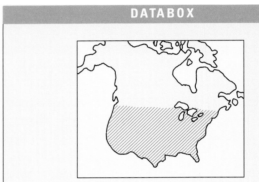

YEAR ROUND

SUMMER

DESCRIPTION: Shades of brown with yellow to cream stripes and black dots along sides; reddish legs. 1½ to 2¼ inches long.
SIGN: Clicking and buzzing of wings in flight.
HABITAT: Grassy areas, often near edge of wooded area.
REPRODUCTION/REARING: Female deposits egg mass into soil, from which larvae burrow upwards to surface.
HABITS: Quick to take flight.
HOW TO ATTRACT: Not a particularly desirable species because of its plant-destroying potential.

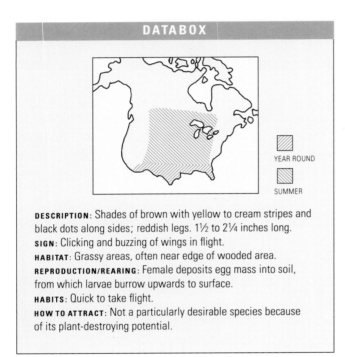

Differential Grasshopper

MELANOPLUS DIFFERENTIALIS

UNLIKE THE AMERICAN BIRD GRASSHOPPER, the differential grasshopper is a homebody. It does not migrate, but overwinters in egg masses buried in the soil. The young grasshoppers emerge from May through July.

The differential grasshopper can be an extremely destructive pest in agricultural crop and fruit-growing areas throughout the United States. It is one of only a half-dozen species of grasshoppers of the 600-plus that have been identified in the U.S. and Canada that account for nearly all agricultural damage done by this family of insects.

DATABOX

DESCRIPTION: Creamy gray-green with distinct black markings between segments and in row of "V"s along hindlegs. 1 to 1¾ inches long.
SIGN: Destruction to plant leaves.
HABITAT: Grasslands and agricultural areas.
REPRODUCTION/REARING: Female lays 8 egg masses of about 10 eggs each into the ground.
HABITS: Often can be observed in dry, dusty spots.
HOW TO ATTRACT: Not a desirable species.

Brown Daddy-long-legs

PHALANGIUM OPILIO

ONE OF 200 SPECIES of this spider-like family in North America, the daddy-long-legs is in fact not a true spider. Also known as the harvestman, it is easily distinguished from spiders by its long legs that give it the appearance of an alien creature.

While daddy-long-legs are predators, feeding on smaller insects as well as decaying animal matter, they cannot inflict bitten injuries on animals larger than themselves. Instead, they rely on potent stink glands and the ability to pull free of any of their legs that may be grasped by a predator.

DATABOX

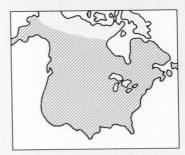

DESCRIPTION: Red-brown body with eight long, thin legs. Body is ⅛ to ¼ inch long.

HABITAT: Grassy areas, often near trees.

REPRODUCTION/REARING: Female lays eggs into soil in the fall. Eggs hatch and young emerge with first warm days of spring. They mature by the end of summer and then mate.

HABITS: Often seen on sides of buildings or trees, sunning itself.

HOW TO ATTRACT: No pesticides.

Ring-legged Earwig

EUBORELLIA ANNULIPES

INFESTATIONS OF THESE SMALL, fearsome-looking insects are a cause of alarm for many a gardener across the United States, but in fact they are as often a beneficial predator as a pest. While the ring-legged earwig is a common sight among the garden plants, which it has been reported to feed upon, it is just as likely hunting for its prey, aphids, and other small insects.

DATABOX

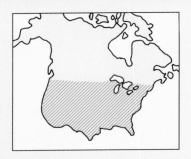

DESCRIPTION: Dark brown with lighter legs and just behind head. ¼ to 1 inch long.
HABITAT: Agricultural areas and gardens.
REPRODUCTION/REARING: Female lays eggs into small holes under leaf litter in the fall and remains with them until they hatch the following spring. The nymphs grow to maturity by the end of summer.
HABITS: Tend to turn up in garden vegetables, such as cabbage, although generally not destructive to the plants.
HOW TO ATTRACT: Plants that attract aphids.

Millipede

GENUS – JULUS
SEVERAL SPECIES

MILLIPEDE AND THOUSAND-LEGS are names commonly applied to any one of more than 50 species of this genus, one or more of which inhabit all of the United States and southern Canada. The common names, however, are an exaggeration; some species have more than 50 segments, each with its own pair of legs.

DATABOX

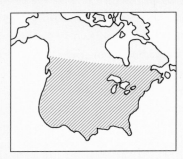

DESCRIPTION: Gray to purplish gray, with as many as 45 body segments, each with its own pair of legs. ½ to ¾ inch long.
HABITAT: Moist meadows and gardens.
REPRODUCTION/REARING: Female lays eggs into damp earth. Young hatch in 3–4 weeks, with 3 body segments and pairs of legs. They undergo many molts as they grow to maturity.
HABITS: Often found in leaf litter and under rotting logs.
HOW TO ATTRACT: Correct habitat, including leaf litter and rotting logs.

Ladybug Beetles

FAMILY – COCCINELLIDA

ALSO KNOWN AS LADYBIRD BEETLES, several species are spread across North America. Most have glossy red to yellow shells with black markings and a black head, and can generally be differentiated by the black markings: for example, two-spotted ladybug beetle, nine-spotted ladybug beetle and three-banded ladybug beetle.

All species are among our most beneficial insects, consuming large quantities of aphids and scale insects. Ecologically-based garden supply houses even see the ladybug beetles as natural aphid controls.

DATABOX

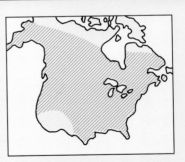

DESCRIPTION: Black head and thorax with cream markings; orange forewings with varying numbers (by species) of black spots. ⅛ to ¼ inch long.

SIGN: Often found on same plants as prey species and aphids.

HABITAT: Moist grassy and weedy areas.

REPRODUCTION/REARING: Female attaches yellowish egg clusters to plant leaves near aphids. Eggs hatch within a few weeks, and larvae feed on aphids and then pupate.

HABITS: Often gather in large numbers at the base of trees in fall.

HOW TO ATTRACT: Plants that attract aphids. Also, sold by environmentally-minded garden supply companies as a natural pest control.

Japanese Beetle

POPILLA JAPONICA

INTRODUCED INTO THE UNITED STATES accidentally in the early 1900s, this native of Japan has become the scourge of gardeners along the entire East Coast. Hundreds of different plants, both wild and domestic, are included in the insect's diet. The adult feeds on the leaves and fruit, while the larva attacks the roots. Tachinid flies have proven to be an important biological control on the Japanese beetle.

DATABOX

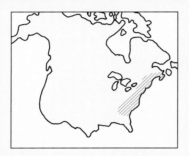

DESCRIPTION: Shiny green with red-brown forewings and patches of white hairs at sides. ½ to 1½ inches long.
SIGN: Destruction to the leaves and fruits of a wide variety of plants.
HABITAT: Open woodlands and grassy areas.
REPRODUCTION/REARING: Female lays several clusters of a few eggs each in the ground. Eggs hatch and larvae are fully grown by fall. They spend the winter in the ground and then pupate in spring.
HABITS: Familiar buzzing, clicking flight.
HOW TO ATTRACT: Not a desirable species.

Northern Walkingstick

DIAPHEROMERA FEMORATA

ALTHOUGH THEY ARE never overly common, there may be more walkingsticks in any given area that one would generally suppose due to their highly effective camouflage. When a walkingstick remains perfectly still, there is generally no distinguishing it from any other stick.

Their large and formidible appearance belies their vegetarian diet, which includes species of nut trees.

DATABOX

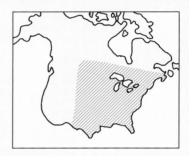

DESCRIPTION: Brown to green-brown, exactly replicating appearance of a stick. 3 to 4 inches long.
SIGN: Even physical presence of the large insect is difficult to discover because of its highly effective camouflage.
HABITAT: Woodlands.
REPRODUCTION/REARING: Female drops eggs individually into leaf litter in the fall, where they will spend the winter months. They hatch in mid-spring and nymphs climb to plant leaves for feeding.
HABITS: Slow-moving. Secretive.
HOW TO ATTRACT: Not readily attracted.

Praying Mantis

MANTIS RELIGIOSA

WITH THE LADYBUG BEETLE, the praying mantis is the gardener's best friend. Nearly all smaller insects are fair game to this large predator, which is lightning fast with its grasping forelegs. It holds the prey crosswise in those forelegs and eats it like a cob of corn.

Although egg cases may contain hundreds of eggs, the praying mantis is a loner except when breeding. If the nymphs are not scattered by the wind when they hatch, they will eat one another until only one remains in a localized territory.

DATABOX

DESCRIPTION: Large, green, stalking insect with bulbous eyes and prayerfully-folded forelegs. 2 to 3½ inches long.

SIGN: Large, brown, hardened egg cases attached to plant stems.

HABITAT: Weedy areas.

REPRODUCTION/REARING: Mates in late summer; female often kills and eats male immediately thereafter. She then attaches the egg case, carrying hundreds of eggs, to a plant stem. The eggs overwinter and miniature replicas of the adult insect hatch in late spring. If the nymphs are not dispersed quickly by the wind, they will immediately begin to eat one another.

HABITS: Adults are loners, except for brief mating.

HOW TO ATTRACT: Do not use pesticides and allow corners of the backyard to become very weedy. Also, egg cases are sold by environmentally-minded garden supply houses.

RIGHT A praying mantis, alighting on goldenrod, is well camouflaged by the surrounding vegetation.

Little Black Ant

MONOMORIUM MINIMUM

THE LITTLE BLACK ANT is the tiny creature that is commonly seen in the house and on the sidewalk, carrying relatively huge bits of food back to its nest. All manner of man's table scraps and garbage is a rich find to this insect, which will soon swarm over the food item.

Nests, which are underground, often are revealed by a small crater at the surface. Several thousand insects and their larvae may be housed below.

Daily observations will reveal that the little black ant follows the same trails every day.

DATABOX

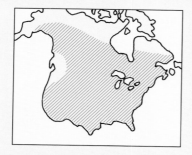

DESCRIPTION: Shiny black with relatively large, "L"-shaped antennae. ⅟₁₆ to ⅛ inch long.
SIGN: Small depressions at entrances to underground nest. Tiny, but regular trails throughout territory.
HABITAT: Open areas.
REPRODUCTION/REARING: Queen begins nest, tending to first brood. Workers from that brood then assume all work.
HABITS: Active round-the-clock, locating and transporting tiny bits of food back to the nest.
HOW TO ATTRACT: Can become a pest if it moves indoors.

Red Ant

GENUS – FORMICA
SEVERAL SPECIES

APHIDS ARE LIKE A MAGNET to many of the species that make up this genus. The ants herd them much like cattle, prodding them occasionally to cause a release of honeydew that is quickly collected and taken back to the nest. Nectar from a large variety of flowers is also collected by the ants themselves.

A particularly widespread species in the east is the Allegheny mound ant, which builds an underground nest with an entrance covered in tiny plant debris.

DATABOX

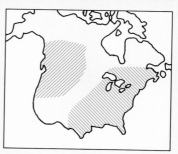

RED ANT

ALLEGHENY MOUND ANT

DESCRIPTION: Orange to red-brown to very dark brown, with darker or black abdomen. ⅛ to ¼ inch long.
SIGN: Entrances to the underground nest are littered with tiny plant debris.
HABITAT: Woodlands.
REPRODUCTION/REARING: Similar to little black ant.
HABITS: Often found with aphids, which they herd like cattle for the sweet honeydew.
HOW TO ATTRACT: Not necessary to try to attract.

Earthworm

LUMBRICUS TERRESTRIS

FEW PEOPLE LIVING in a temperate climate any-where on Earth have not encountered the earth-worm. It is one of the most common animals in the backyard, so long as that backyard has not been "nuked" with pesticides.

It has been estimated that a single acre of soil may contain more than 50,000 of these tiny earth-moving machines. Their tunnelling actions are highly beneficial to the soil, as they bring the dug soil to the surface and "spit" it out, helping to create a new top-layer of fertility.

Heavy rains will flood the earthworm's tunnels, forc-ing them to the surface. Vibrations, such as driving a post into the ground, have also been known to bring them out. They also make nightly excursions to the world above to feed on decaying organic matter, hence the name of nightcrawler.

Many tiny bristles on the underside, combined with the mucus that they secrete, provide the animal's means of movement. Earthworms are not insects, as commonly thought, but are among the 7,000-plus species that make up the Phylum Annelida.

DATABOX

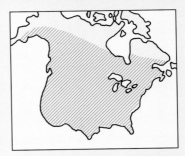

DESCRIPTION: Pinkish-brown to brown. As long as 11 inches.

SIGN: Small holes in ground, with tiny piles of mud about them, mucus trails through grass.

HABITAT: Widely varied.

REPRODUCTION/REARING: Individuals couple in head-to-tail fashion and place sperm in each other's receptacles. Each then secretes a light cocoon over its head, carrying sperm and eggs. Eggs hatch in about 3 weeks.

HABITS: Highly nocturnal. Generally emerge during and after spring and summer rains.

HOW TO ATTRACT: No pesticides.

American Robin

TURDUS MIGRATORIUS

FEW BACKYARD ANIMALS are as instantly recognizable to as many people as is the orange-breasted robin. The "orphaned" chicks are among the first "can I keep him?" creatures captured by many a child. In fact, these birds are rarely orphaned and should be left where they are found for the nearby parent to attend to.

Equally well-known is the robin's hunger for earthworms, which common knowledge tells us are discovered through the bird's sharp hearing – hence the reason for the constant twisting and angling of the head in a hunting robin. However, the bird eats many types of insects, as well as fruit and berries. And it finds earthworms by sight; the twisting of the head is for an improved field of vision.

Another common myth about this myth-surrounded species is that the robin is a harbinger of spring, returning from a southern migration. In fact, the American robin often spends the winter in the forests of the north, moving back to the lawns and gardens as spring replaces winter.

DATABOX

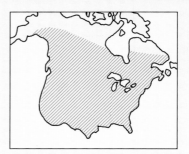

DESCRIPTION: Gray across head, back, and wings, darker at head; orange-red breast. 9 to 11 inches long.
SIGN: Voice is a "cheerup, cheerio, cheerup."
HABITAT: Open, grassy areas with some trees.
REPRODUCTION/REARING: Female lays 3–5 blue eggs in cup-nest of mud and grass, lined with softer plant fibers. Incubation is often by both parents. Eggs hatch in 12–14 days. 2–3 broods per year.
HABITS: Spends a great deal of the day hunting through grassy areas.
HOW TO ATTRACT: Suitable habitat.

European Starling

STURNUS VULGARIS

A DESPISED, NON-NATIVE species, the starling is nonetheless very successful. From a few dozen released into New York City in the late 1800s, the bird has spread across the continent in huge numbers. In fact, fall roosts may contain millions of the birds.

The starling is an extremely aggressive bird, robbing native species, such as the bluebird, of their nesting cavities and hogging bird feeders. It also is not the least bit finicky about what it eats, including everything from insects to grain to seeds to berries in its diet.

Showing a preference for the company of man and the easy pickings he offers to an animal of such diverse eating habits, the starling shuns forested areas for agricultural and residential surroundings. From a gardener's point of view, the starling may actually be a beneficial species because of its taste for insects.

The fall and winter plumage is an unimpressive, dull brown-black, covered with white spots. But, in the right light in spring and summer, the starling sports a surprisingly attractive purplish-green sheen.

DATABOX

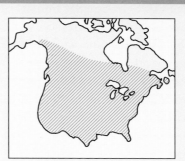

DESCRIPTION: Shiny black with white flecks; long, yellow bill. 7½ to 8½ inches long.

SIGN: Voice is a "whee-ee" or mimickry of other birds.

HABITAT: Residential and agricultural areas.

REPRODUCTION/REARING: Female lays 4–6 blue eggs in nest of twigs and debris inside a cavity. Both parents incubate. Eggs hatch in 11–14 days. 1–3 broods per year.

HABITS: Gathers in flocks, except during mating season, wherever food is available.

HOW TO ATTRACT: No need or reason for special effort. Starlings will show up for whatever is offered for other animals.

Northern Mockingbird

MIMUS POLYGLOTTOS

THE FIRST PART of this bird's name was something of a misnomer in past decades. The northern mockingbird was a resident of the southern United States until early in the 20th century, when it began a northward expansion of its territory. It is now a year-round resident as far north as Maine and the northern Midwest. The clearing by man of much of the forests in the east and the planting of many ornamental, berry-producing shrubs and vines are thought to be principal reasons for the expansion.

The second part of the name, however, could not be more true. The northern mockingbird is able to duplicate the calls of any bird it encounters, as well as many other non-avian sounds, such as human whistles. It can combine these songs into almost non-stop strings, flowing smoothly from one to the other.

The northern mockingbird is a highly territorial species, generally intolerant of all other birds its size. It tolerates a mate and the growing offspring in the spring and summer only until the young ones can fend for themselves.

DATABOX

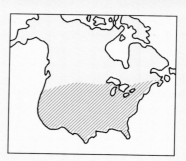

DESCRIPTION: Gray with white patches on wings and on long tail; lighter underside. 9 to 11 inches long.

SIGN: Voice is mimickry of other birds, often in long strings.

HABITAT: Open areas with thickets or hedgerows nearby.

REPRODUCTION/REARING: Female lays 3–5 blue-green eggs with brown spots into cup-nest of sticks and weed stems, generally placed in small, thorny shrub. Female incubates eggs for 12–14 days. 2–3 broods per year.

HABITS: Very territorial. Patrols home territory with great frequency.

HOW TO ATTRACT: Thickets and hedgerows. Suet, fruit, and nut meats at feeder.

White-throated Sparrow

ZONOTRICHIA ALBICOLLIS

A COMMON WINTER RESIDENT of much of the United States, flocks of white-throated sparrows are easily attracted to and maintained at backyard feeders. But as spring returns, the sparrows move back to Canada to breed.

They prefer a wooded or shrub-intense backyard, particularly coniferous plants, and are often found in forest thickets and stream banks, where berry-producing plants are abundant.

The white-throated sparrow female produces three to five pale green, spotted brown eggs in a cup of grass and moss on the ground or nearly at ground level in a small shrub.

DATABOX

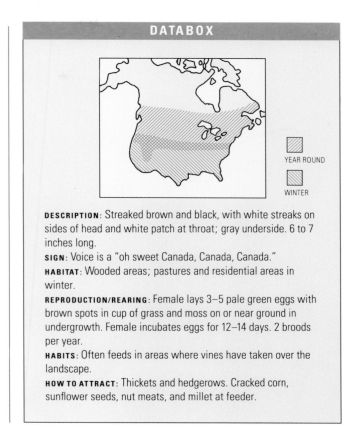

YEAR ROUND

WINTER

DESCRIPTION: Streaked brown and black, with white streaks on sides of head and white patch at throat; gray underside. 6 to 7 inches long.

SIGN: Voice is a "oh sweet Canada, Canada, Canada."

HABITAT: Wooded areas; pastures and residential areas in winter.

REPRODUCTION/REARING: Female lays 3–5 pale green eggs with brown spots in cup of grass and moss on or near ground in undergrowth. Female incubates eggs for 12–14 days. 2 broods per year.

HABITS: Often feeds in areas where vines have taken over the landscape.

HOW TO ATTRACT: Thickets and hedgerows. Cracked corn, sunflower seeds, nut meats, and millet at feeder.

House Finch

CARPODACUS MEXICANUS

THE HOUSE FINCH is not entirely an invading, non-resident of North America. But its presence in the East is due to its introduction as a cage bird that was set free in the 1940s by New York City pet shop owners trying to avoid penalties for selling a native species. Since that time it has become one of the more common backyard birds in the East.

The numbers and range continue to expand for this prolific species, which can raise two or three broods of six chicks each every year.

In its native West, the house finch is a species of agricultural lands. But in the East, true to its invader status, the bird is tied closely to man. Flocks of 100 or more are common at winter feeders throughout the region.

Nesting sites include backyard trees and shrubs, birdboxes, and frequently man's hanging planters, where it will scoop out a thin depression amid the domestic plants.

DATABOX

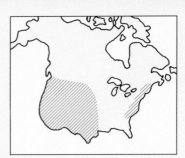

DESCRIPTION: Male is streaked brown and buff, with a purplish red breast, forehead and rump. Female lacks the red. 5 to 6 inches long.

SIGN: Voice is a warble.

HABITAT: Residential areas.

REPRODUCTION/REARING: Female lays 2–6 pale blue eggs spotted with black in cup of grass in cavity or thicket. Female, primarily, incubates for 12–14 days. 2 broods per year.

HABITS: Gathers quickly in large flocks at backyard feeders.

HOW TO ATTRACT: No special efforts needed; will take advantage of whatever is offered for other animals.

Northern Cardinal

CARDINALIS CARDINALIS

A WELCOME VISITOR to any winter feeder, the male northern cardinal brings a bright flash of red to the sometimes dingy winter setting. The female is a much duller bird, showing only a touch of red at the wings, tail and crest.

Both can be somewhat shy birds. Seemingly aware of the potential to attract predators with its bright coloring and easily bullied, the cardinal prefers a feeder that is close to thick shrubs or evergreen trees. Both sexes, however, will defend their year-round home territory from others of the same sex.

The cardinal has been extending its range northward in recent decades, taking advantage of suburban environments that provide plenty of the needed cover and ample food sources.

ABOVE The bright red coloring of the male northern cardinal is unmistakable, particularly in a winter setting.

DATABOX

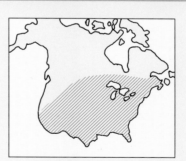

DESCRIPTION: Male is bright red with black face and red beak. Female is buff with a red hint on wings, rest, and tail. 8 to 9 inches long.

SIGN: Voice is "cheer, cheer, cheer, whoit, whoit, whoit."

HABITAT: Residential areas and woodland edges with thickets and evergreen trees.

REPRODUCTION/REARING: Female lays 2–5 pale green eggs spotted brown in cup of twigs and plant fibers, usually placed in thicket. Female incubates eggs for 12 days. 1–3 broods per year.

HABITS: A shy bird that prefers to stay close to cover.

HOW TO ATTRACT: Thickets, evergreen trees; feeders placed near cover; sunflower seeds and raisins at feeder.

American Goldfinch

CARDUELIS TRISTIS

THE AMERICAN GOLDFINCH goes by several other, highly accurate names that are helpful in describing the bird and its habits. The bright lemon-yellow male of spring and summer certainly deserves the name of wild canary. In the late fall and winter it takes on the same dull olive coloring of the female.

And, the name of thistle finch certainly reminds us of the bird's fondness for that plant. While the seeds are a primary diet item, the goldfinch also lines its nest with the fluffy thistledown from the plant.

Common throughout the entire continent, except for the far north, during the warmer months, the goldfinch generally retreats south of the Canadian-U.S. border for the winter. It will become a regular feeder visitor to any hanging feeder that is stocked with thistle, or niger, seeds.

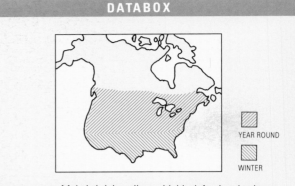

DATABOX

YEAR ROUND

WINTER

DESCRIPTION: Male is bright yellow with black forehead, wings and tail, and white rump, spring through early fall. Female, and male in fall through early spring, is olive-gray with black tail and wings, and white wing bars. 4 to 5½ inches long.
SIGN: Voice is a "per-chickory, per-chickory."
HABITAT: Residential areas and grasslands with trees and thickets.
REPRODUCTION/REARING: Female lays 4–6 blue eggs in a cup of grass bark, and plant fibers, lined with thistledown, in a tree or shrub. Female incubates eggs for 11–14 days. 1 brood per year.
HABITS: Life is tied closely to availability of thistles.
HOW TO ATTRACT: Thistle (niger) seed at feeder.

RIGHT A male American goldfinch displays typical fall to early spring coloration – bright yellow with black markings on head, wings and tail.

Evening Grosbeak

COCCOTHRAUSTES VESPERTINUS

DEROGATORALLY REFERRED TO as the gross pig by some backyard feeder enthusiasts for the incredible amounts of choice sunflower seeds that a flock of these birds will eat every day, the evening grosbeak is nonetheless a common backyard bird during the winter.

However, the bird's seasonal movements are so erratic as to create a system of peaks and valleys in the southern wintering range. Some years the evening grosbeak will be abundant. Other years it will be scarce or absent. When the cone crop is abundant in their northern breeding grounds, the birds will winter there. When the cone crop slacks off about every two or three years, they wander to the southeast in search of more ample food supplies.

Prior to the boom in backyard bird feeding, the grosbeak bred no further east than the western Great Lakes region. But the availability of this man-provided winter food source has helped the bird to extend its range to the East Coast.

DATABOX

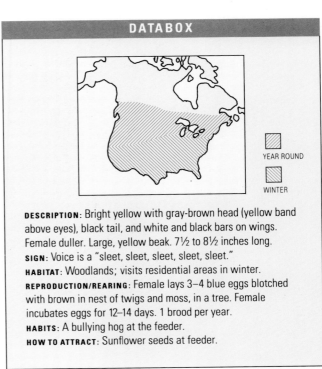

YEAR ROUND

WINTER

DESCRIPTION: Bright yellow with gray-brown head (yellow band above eyes), black tail, and white and black bars on wings. Female duller. Large, yellow beak. 7½ to 8½ inches long.
SIGN: Voice is a "sleet, sleet, sleet, sleet, sleet."
HABITAT: Woodlands; visits residential areas in winter.
REPRODUCTION/REARING: Female lays 3–4 blue eggs blotched with brown in nest of twigs and moss, in a tree. Female incubates eggs for 12–14 days. 1 brood per year.
HABITS: A bullying hog at the feeder.
HOW TO ATTRACT: Sunflower seeds at feeder.

Song Sparrow

MELOSPIZA MELODIA

THE WIDESPREAD SONG SPARROW is a highly variable species, with more than 30 subspecies forming the total population that covers most of the continent. The subspecies are quite varying in both appearance, size, and song patterns.

Research has determined that the song sparrow is hatched with an instinct to sing its basic song, which it then modifies as it learns the sound of other song sparrows in its immediate vicinity. Through this process, all the birds from one region may include patterns in their songs that are lacking or at least quite different in the songs of birds from other regions.

True to their name, song sparrows are near-constant singers, letting loose with their voices at any time of day and in all seasons. Their warning calls are significantly different from their normal song and vary according to the location and type of perceived threat. One signal warns of an approaching hawk, another of danger on the ground (such as a cat), and a third of cowbirds invading the territory. The cowbird call is significant to the song sparrow, as the parasitic egg-layer uses it regularly to hatch and rear its young.

DATABOX

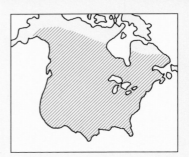

DESCRIPTION: Streaked in brown, black, gray, and white, with distinct brown stripes on crown and gray to off-white underside streaked with dark brown. 5 to 7 inches long.

SIGN: Voice varies widely by population, but generally includes three or four whistles, followed by a buzz, followed by a trill.

HABITAT: Residential areas, thickets, and pastures.

REPRODUCTION/REARING: Female lays 3–5 green eggs spotted with brown in nest of grass and leaves, lined with hair, on or near the ground. Female incubates eggs for 12–15 days. 2–3 broods per year.

HABITS: A prolific singer.

HOW TO ATTRACT: Sunflower seeds, cracked corn, suet, and peanut butter at feeder.

Barn Swallow

HIRUNDO RUSTICA

AGRICULTURAL AREAS IN practically any environment on the continent are home to the barn swallow, which also includes rural and suburban areas in its territory. It is an extremely beneficial species, eating hundreds of flying insects every day. It has been estimated that a single bird may fly more than 500 miles per day in pursuit of insects, when it is feeding a nestful of young.

The barn swallow builds its nest of straw and grass cemented together with mud that it has collected in hundreds of trips to nearby mud puddles on any flat area that offers shelter. Barns were the preferred location in more agricultural times, hence the bird's name.

Nests will sometimes be built in colonies of dozens of pairs of birds, but this is the exception rather than the rule. The barn swallow is not nearly as colonial as other swallows.

Barn swallows are migratory birds, wintering in South America.

DATABOX

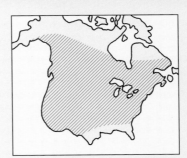

DESCRIPTION: Shiny blue to blue-black with red-brown at face; buff underside; long, forked tail. 7 inches long.

SIGN: Voice is a non-stop chirping and twittering.

HABITAT: Agricultural areas and lawns, generally near water.

REPRODUCTION/REARING: Female lays 3–6 white eggs spotted with brown into mud and straw nest, often placed in man's outbuildings. Both parents incubate eggs for 14–16 days. 2–3 broods per year.

HABITS: Active dawn to dusk, hunting insects on the wing.

HOW TO ATTRACT: Promotion of flying insects.

House Sparrow

PASSER DOMESTICUS

EVERYWHERE THAT HUMANS are found, so is the house sparrow. The small, non-native bird was introduced to North America, beginning in the mid-1800s, in various failed attempts to use it as a biological control on agricultural pests.

It may have failed those tests, but it found and exploited a niche that no native bird was occupying. That niche is in close proximity to man, making use of the manmade environment and the discards of human society. City parks and suburban lawns provided abundant fare to the invader and it spread across the continent.

Similar tales accompany the small bird's introduction from Europe and Africa to nearly every temperate location on Earth.

Like other invaders, the house sparrows competes vigorously and successfully with native birds for nesting cavities. For much of the 20th century this helped to cause a decline in species such as the bluebird, until man's intervention with additional nesting sites came to the native bird's rescue. The house sparrow is not choosey about its nesting spot and will accept near any sheltered spot to hatch and raise its young.

DATABOX

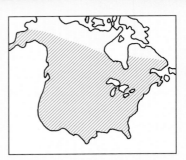

DESCRIPTION: Brown-gray with buff underside. Male has black throat, white cheeks, and brown nape. 5 to 6½ inches long.
SIGN: Voice is a constant, deadpan chirping.
HABITAT: Near man.
REPRODUCTION/REARING: Female lays 3–7 white eggs speckled with brown in nest of loose grass and debris in a cavity. Female incubates eggs for 11–12 days. 3 broods per year.
HABITS: Relatively tame.
HOW TO ATTRACT: No special effort required; will take advantage of any offering.

ABOVE The two male house sparrows in the foreground are distinguished from the female of the species by their black throats.

Blue Jay

CYANOCITTA CRISTATA

FROWNED UPON AS a backyard species by many bird-feeding enthusiasts because of its bullying and hogging of the choicest seeds, the blue jay is welcomed by just as many for the brilliant color it introduced into the winter backyard and for its showy ways.

The blue jay approaches the feeder in a flourishing swoop, screaming out a call as it approaches. It selects the larger seeds and nuts – sunflower seeds and peanuts are favorites – to steal away from the feeder. It may break some open and eat them at a nearby perch, but it will hide just as many in nearby leaf litter and the crooks of tree branches.

It accepts no bullying from other birds while at the feeder, where it generally will remain only until the seeds of its choosing are depleted. The blue jay can be expected to appear at nearly the same time each morning if the food supply is maintained.

DATABOX

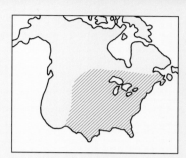

DESCRIPTION: Bright blue with white and black in wings, and black at crest and face; white underside. 12 inches long.
SIGN: Voice is "jaaaay, jaaaay, jaaaay."
HABITAT: Open areas near woodlands.
REPRODUCTION/REARING: Female lays 3–6 olive eggs spotted with brown in bowl of sticks and leaves in the fork of a tree. Both parents incubate eggs for 15–17 days. 1–2 broods per year.
HABITS: A bully at the feeder. Appears at nearly the same time each morning.
HOW TO ATTRACT: Suet, whole peanuts, acorns and other nuts, and sunflower seeds at the feeder.

LEFT The blue jay will not be dislodged from a feeder by other species, especially if sunflower seeds are on offer.

White-breasted Nuthatch

SITTA CAROLINENSIS

NICKNAMED THE UPSIDE-DOWN BIRD, the white-breasted nuthatch certainly has earned that name. It typically will be seen descending the side of a tree, with its black-capped head pointing down.

This unusual habit is a feeding adaptation, allowing the nuthatch to observe and snatch insects in the bark that other tree-clinging birds, such as the downy woodpecker, may have missed in its more conventional right-side-up approach.

The white-breasted nuthatch generally travels with its mate year-round, although the pair will join the mixed winter flocks that also include woodpeckers, creepers, kinglets, and chickadees, while the flock is in the nuthatches' home territory.

Sunflower seeds are a favorite feeder food, but a white-breasted nuthatch with a larger item – such as a peanut or an acorn – provides much more entertainment. The small bird will wedge the nut into a crevice in a tree and hammer away at it with its strong beak until breaking through to the meat inside.

DATABOX

DESCRIPTION: Blue-gray with black cap, white face and underside. 5–6 inches long.
SIGN: Voice is a "yank, yank, yank."
HABITAT: Mixed forests.
REPRODUCTION/REARING: Female lays 5–10 white eggs speckled brown in nest of grass and twigs, lined with hair and plant fibers, in cavity. Female incubates eggs for 12–14 days. 1 brood per year.
HABITS: Generally seen moving down tree trunk head-first, hunting for insects in the bark.
HOW TO ATTRACT: Correct habitat; suet, raw beef, nut meats, and sunflower seeds at feeder.

Black-capped Chickadee

PARUS ATRICAPILLUS

A COMMON, RELATIVELY tame visitor to the bird feeder, the black-capped chickadee is also easily trained to take food from a human's hand. Patience is the watchword, as the would-be hand-feeder must move gradually closer to the feeder over several days – don't expect success at the first attempt. Standing very still and yet relaxed, hold out a cupped hand filled with sunflower seeds. Soon the small birds will be perching on your fingers to take the seeds.

Timing each day is made somewhat easier by the relative regularity with which the mixed flocks that include chickadees appear at the feeder, as they make their daily rounds of food source. These mixed flocks are generally comprised of nuthatches, creepers, kinglets, woodpeckers, and chickadees.

During the winter, while at the feeder the chickadee will travel in flocks, but these groups break up for the breeding season each spring.

DATABOX

YEAR ROUND

WINTER

DESCRIPTION: Gray with black cap and throat, white cheeks and white to off-white underside. 4½ to 5¾ inches long.
SIGN: Voice is a "chick-a-dee-dee-dee."
HABITAT: Woodlands; residential areas in winter months.
REPRODUCTION/REARING: Female lays 5–10 white eggs spotted brown in cup of grass, moss and plant down in a cavity. Both parents incubate eggs for 11–13 days. 1–2 broods per year.
HABITS: Easily tamed.
HOW TO ATTRACT: Sunflower seeds at feeder.

Mourning Dove

ZENAIDA MACROURA

PRIOR TO THE COMING of the Europeans to North America, the open-land-loving mourning dove was a much less common bird. Today, however, it has taken advantage of the massive opening of the continent for agricultural lands and is one of our most common species. It continues its range expansion each year.

Large numbers of mourning doves will quickly become nearly permanent residents in any backyard that offers them cracked corn scattered on the ground. City parks will also support large flocks of the birds.

In many states the bird is considered a game species and hunted in great numbers, providing something of a balancing effect on the population. In others, where it is classified as a songbird and protected from hunting, the mourning dove has become as much of a pest as the common pigeon.

ABOVE A mourning dove shelters its chick in a nest of twigs.

DATABOX

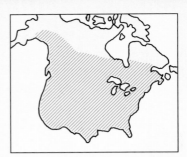

DESCRIPTION: Gray-brown with black spots on wings, cream head with black eye, cream underside. 12 inches long.
SIGN: Voice is a throaty, rolling "cooooo, cooooo."
HABITAT: Agricultural and residential areas.
REPRODUCTION/REARING: Female lays 2–4 white eggs, in nest of twigs in tree or shrub. Both parents incubate eggs for 13–15 days. 2 or more broods per year.
HABITS: Will gather in gradually growing numbers at food source.
HOW TO ATTRACT: Cracked corn on the ground.

Downy Woodpecker

PICOIDES PUBESCENS

MEASURING LESS THAN 7 inches in length, the downy woodpecker is North America's smallest woodpecker. It is also the most common, generally found across the continent, except for desert and far north areas. And, finally, it is the tamest of our woodpeckers, easily attracted into the backyard by an offering of suet.

Like other woodpeckers, the downy woodpecker is well-designed for drilling its way into tree bark in search of the insects that are hidden there. The short, straight beak is extremely hard and its impacts are cushioned at the base of the skull by a padding muscle. The long tongue ends in barbs that act exactly like the barbs on a fish-hook in snagging and grasping the insects. Long claws and strong tail feathers provide support while the bird hammers at the tree.

Also like other woodpeckers, the male downy woodpecker sets about a very rapid tapping in the spring that serves the same purpose as springtime song in many other male birds – to attract a mate and to warn off other male downy woodpeckers.

This little woodpecker is commonly seen among the mixed flocks of winter that also include the chickadees, creepers, nuthatches, and kinglets.

DATABOX

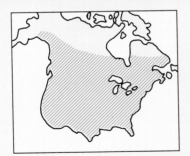

DESCRIPTION: Black with bands of white patches; buff to gray underside. Male has red patch on nape. 6 to 7 inches long.
SIGN: Rapid tapping on tree trunk. Voice is a "pick."
HABITAT: Woodlands, wooded residential areas, wooded parks.
REPRODUCTION/REARING: Female lays 3–6 white eggs atop sawdust in a cavity. Both parents incubate eggs for 12 days. 1 brood per year.
HABITS: Most often seen clinging to and scaling side of tree in search of insects.
HOW TO ATTRACT: Correct habitat; suet, peanut butter, and sunflower seeds at feeder.

House Wren

TROGLODYTES AEDON

THE HOUSE WREN selects some of the oddest locations for its nest of any North American birds. Shirts hanging on the laundry line, old hats, mailboxes, drain pipes, and discarded tin cans have all held the tiny bird's nest at one time or another. It should then come as no surprise that the house wren will readily accept nesting boxes.

The bird is also extremely loyal to its nesting grounds, returning there after a winter spent in the extreme south of the continent. If its nest has survived from the previous year, it will rip it apart and rebuild, incorporating some of the same discarded materials back into the new nest.

While that may all be quite endearing, the house wren has one habit that is not so lovable. It systematically seeks out the nests of other birds in its small home territory and destroys their eggs by piercing them.

BELOW A house wren, holding a twig in its beak, perches on a birch tree.

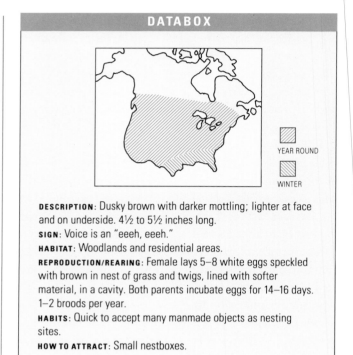

DATABOX

YEAR ROUND

WINTER

DESCRIPTION: Dusky brown with darker mottling; lighter at face and on underside. 4½ to 5½ inches long.
SIGN: Voice is an "eeeh, eeeh."
HABITAT: Woodlands and residential areas.
REPRODUCTION/REARING: Female lays 5–8 white eggs speckled with brown in nest of grass and twigs, lined with softer material, in a cavity. Both parents incubate eggs for 14–16 days. 1–2 broods per year.
HABITS: Quick to accept many manmade objects as nesting sites.
HOW TO ATTRACT: Small nestboxes.

Common Grackle

QUISALUS QUISCULA

OFTEN INCORRECTLY REFERRED to as a black-bird, the common grackle actually sports an attractive irridescence over its totally black plumage. There are actually a bronze and purple phases among the males, and in the right light they can even have a greenish sheen. Two bright yellow eyes finish the "look."

This most abundant and smallest of North America's grackle is an extremely scrappy bird among its own kind. The territorial threat entails the puffing out of the feathers about the neck, an arching of the neck toward the target and a gradual "yawning" with the wings angled over the back. Direct contact is common.

Insects, fruits, and berries make up the bulk of the common grackle's diet. But it is a highly opportunistic bird. It will spend great amounts of time at the feeder and will kill and eat almost any smaller animal that it can catch. Bird eggs and chicks are never safe from this species.

DATABOX

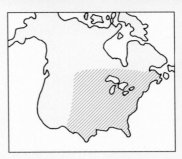

DESCRIPTION: Black with an iridescence of blue to bronze; bright yellow eyes. 11 to 13½ inches long.

HABITAT: Residential areas, agricultural areas, and woodlands.

REPRODUCTION/REARING: Female lays 3–6 blue eggs spotted brown in mass of twigs lined with grass, in a tree. Female incubates eggs for 14–15 days. 1 brood per year.

HABITS: A large, scrappy bird among its own kind.

HOW TO ATTRACT: Sunflower seeds, and cracked corn at the feeder. Table scraps and bread crumbs scattered on the lawn.

Red-winged Blackbird

AGELAIUS PHOENICEUS

OVER THE PAST THREE DECADES the red-winged blackbird has become a much more common resident of a wider portion of the North American continent, as it has adapted to new types of nesting habitat. Historically restricted to the marshes and other wetlands, the red-winged blackbird of today is just as likely to nest in agricultural fields and pastures, and large backyards. Its one criteria for all habitat that it uses is that it provide heavy cover for the nests.

Because of its after-breeding habit of gathering with other blackbirds in flocks that can range into the millions, the red-winged blackbird is sometimes considered a pest and targeted for eradication efforts. Producing two or three clutches of fledglings each year, the red-winged blackbird springs back quite readily.

The male of this species is unmistakably identifiable from its name. It is a shiny blackbird that sports bright red shoulder bands. The female and the young are less recognizable for their name; they are mottled in dull brown and lack the shoulder bands.

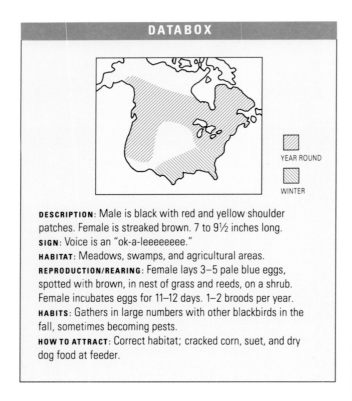

DATABOX

YEAR ROUND

WINTER

DESCRIPTION: Male is black with red and yellow shoulder patches. Female is streaked brown. 7 to 9½ inches long.
SIGN: Voice is an "ok-a-leeeeeeee."
HABITAT: Meadows, swamps, and agricultural areas.
REPRODUCTION/REARING: Female lays 3–5 pale blue eggs, spotted with brown, in nest of grass and reeds, on a shrub. Female incubates eggs for 11–12 days. 1–2 broods per year.
HABITS: Gathers in large numbers with other blackbirds in the fall, sometimes becoming pests.
HOW TO ATTRACT: Correct habitat; cracked corn, suet, and dry dog food at feeder.

RIGHT A red-winged blackbird displays its characteristic red and yellow shoulder patches.

Brown-headed Cowbird

MOLOTHRUS ATER

THE BROWN-HEADED COWBIRD is what is known as a brood parasite. Like the closely related bronzed cowbird, it lays its eggs into the nests of other birds, forcing the other birds to become unknowing foster parents. The two species of cowbird are the only brood parasitic birds in North America.

Many smaller bird species are the target of the cowbird's carefree attitude toward parenting. More than 200 different species have been recorded.

Most will simply raise the cowbird's offspring at the expense of their own. The young cowbird is a rapid grower and will soon push the parent birds' real offspring out of the nest to their death or at least hog most of the food that the parent birds bring back to the nest.

The name cowbird was given to this genus of songbird for a different species that frequented cow pastures, snatching parasitic insects from the mammals' backs. In earlier times, when bison herds of several million animals roamed the North American continent, the brown-headed cowbird followed this same habit with those herds. Its common name in the Old West was buffalo bird.

DATABOX

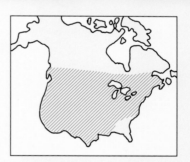

DESCRIPTION: Black with glossy brown head; female is duller.

SIGN: Voice is a rattling gurgle.

HABITAT: Agricultural and residential areas.

REPRODUCTION/REARING: Female lays 4–5 white eggs speckled brown into the nests of other birds (1 per nest) for the unwitting foster parents to hatch and raise as their own.

HABITS: Flocks of this species are quick to gather at any food source.

HOW TO ATTRACT: Not a desirable species, due to its brood-parasitic ways.

ABOVE A male brown-headed cowbird alights in a patch of wild violets.

INDEX

A

Acheta domestica (House cricket), 65
Acris crepitans (Northern cricket frog), 43
Actias luna (Lunar moth), 59
Agelaius phoeniceus (Red-winged blackbird), 92
Agelenopsis genus (Grass spider), 63
Alfalfa butterfly *see* Orange sulphur buttefly (*Colias eurytheme*)
Ambystoma,
    maculatum (Spotted salamander), 39
    tigrinum (Tiger salamander), 39
Anasa tristis (Squash bug), 61
Anole, green (*Anolis carolinensis*), 46
Anolis carolinensis (Green anole), 46
Ant,
    little black (*Monomorium minimum*), 72
    red (genus – *Formica*), 72
Apis mellifera (Honey bee), 53
Araneidae (Orb-weaving spiders), 64
Araneus diadematus (Garden spider), 64

B

Backyards, 7–11
Bat,
    big brown (*Eptesicus fuscus*), 24
    little brown (*Myotis lucifugus*), 23
Bear, black (*Ursus americanus*), 8, 32
Bee,
    American bumble (*Bombus pennsylvanicus*), 53
    honey (*Apis mellifera*), 53
Beetle,
    Colorado potato (*Leptinotarsa decimlineata*), 62
    Japanese (*Popilla japonica*), 70
    ladybug (family – *Coccinellida*), 69
    striped blister (*Epicauta vittata*), 31
Blackbird, red-winged (*Agelaius phoeniceus*), 92
Blarina brevicauda (Short-tailed shrew), 21
Bombus pennsylvanicus (American bumble bee), 53
Bufo americanus (American toad), 44

Bug,
    potato *see* Colorado potato beetle
    small eastern milkweed (*Lygaeus kalmii*), 62
    squash (*Anasa tristis*), 61
Butterflies,
    alfalfa *see* orange sulphur
    American copper (*Lycaena phlaeas*), 56
    cabbage white (*Pieris rapae*), 54
    monarch (*Danaus plexippus*), 55, 62
    mourning cloak (*Nymphalis antiopa*), 55

orange sulphur (*Colias eurytheme*), 54
red admiral, *8*
silver-spotted skipper (*Epargyreus clarus*), 57
tiger swallowtail (*Papilio glaucus*), 56

C

Canis latrans (Coyote), 30
Cardinal, northern (*Cardinalis cardinalis*), 79
Cardinalis cardinalis, (Northern cardinal) 79
Carduelis tristis (American goldfinch), 80
Carprodacus mexicanus (House finch), 78
Chameleon, 46
Chickadee, black-capped (*Parus atricapillus*), 87
Chipmunk, *8*
    eastern (*Tamias striatus*), 12
Chrysemys picta (Painted turtle), 45
Coccinellida family (Ladybug beetles), 69
Coccothraustes vespertinus (Evening grosbeak), 81
Colias eurytheme (Orange sulphur butterfly), 54
Condylura cristata (Star-nosed mole), 22
Cowbird, brown-headed (*Molothrus ater*), 93
Coyote (*Canis latrans*), 30
Cricket, 7
    field (*Gryllus pennsylvanicus*), 65
    house (*Acheta domestica*), 65
Cyanocitta cristata (Blue jay), 85

D

Daddy-long-legs, brown (*Phalangium opilio*), 67
Damselflies (*Ordonata*), 60
Danaus plexippus (Monarch butterfly), 55, 62
Deer, white-tailed (*Odocoileus virginianus*), 6, 36
Diadophis punctatus (Ringneck snake), 48
Diapheromera femorata (Northern walkingstick), 70
Didelphis virginiana (Virgina opossum), 25
Dove, mourning (*Zenaida macroura*), 88
Dragonflies and damselflies (order – *Ordonata*), 60

E

Earthworm (*Lumbricus terrestris*), 73
Earwig, ring-legged (*Euborellia annulipes*), 68
Eastern cottontail (*Sylvilagus floridanus*), 19
Echolocation, 24
Elaphe obsoleta (Rat snake), 48
Epargyreus clarus (Silver-spotted skipper butterfly), 57

Epicauta vittata (Striped blister beetle), 61
Eptesicus fuscus (Big brown bat), 24
Erethizon dorsatum (Porcupine), 29
Euborellia annulipes (Ring-legged earwig), 68
Eumeces fasciatus (Five-lined skink), 47
Evening grosbeak (*Coccothraustes vespertinus*), 81

F

Finch, house (*Carprodacus mexicanus*), 78
    thistle *see* goldfinch, American
Firefly, Pennsylvania (*Photuris pennsylvanicus*), 60
Formica – genus (Red ant), 72
Fox, red (*Vulpes vulpes*), 31
Frogs, 40–43
    bullfrog (*Rana catesbeiana*), 40
    chorus frog (*Pseudacris triseriata*), 43
    common gray treefrog (*Hyla versicolor*), 42
    green (*Rana clamitans*), 40
    northern cricket frog (*Acris crepitans*), 43
    pickerel (*Rana palustrus*), 41
    spring peeper (*Hyla crucifer*), 42
    treefrogs, 42
    woodfrog (*Rana sylvatica*), 41

G

Glaucomys,
    sabrinus (Northern flying squirrel), 16
    volanus (Southern flying squirrel), 16
Goldfinch, American (*Carduelis tristis*), 80
Grackle, common (*Quisalus quiscula*), 91
Grasshopper,
    American bird (*Schistocerca americana*), 66
    differential (*Melanoplus differentialis*), 66
Groundhog *see* Woodchuck
Gryllus pennsylvanicus (Field cricket), 65

H

Hare *see* Jack rabbit
Harvestman *see* Daddy-long-legs
Hermaris thysbe (Clear-winged sphinx moth), 58
Hetorodon,
    nasicus (Western hognose snake), 49
    platyrhinos (Eastern hognose snake), 49
Hirundo rustica (Barn swallow), 83
Hornet,
    bald-faced (*Vespula maculata*), 52
    yellow jacket,
        eastern – *Vespula maculifrons*
        western – *Vespula pennsylvanica*, 52
Hyalophora cecropia (Cecropia moth), 59

Hyla,
    crucifer (Spring peeper), 42
    versicolor (Common gray treefrog), 42
Hyles leneata (White-lined sphinx moth), 58

I

Isia Isabella (Isabella tiger moth), 58

J

Jack rabbit,
    black tailed – *Lepus californicus*
    white tailed – *Lepus townsendii*, 20
Jay, blue (*Cyanocitta cristata*), 85
Julus – genus (Millipede), 68

K

Katydid, true (*Pterophylla camellifolia*), 65

L

Ladybird beetles *see* Beetle, ladybug
Lampropeltis triangulum (Milk snake), 49
Leptinotarsa decimlineata (Colorado potato beetle), 62
Lepus,
    californicus (Black-tailed jack rabbit), 20
    townsendii (White-tailed jack rabbit), 20
Lizard,
    eastern fence (*Sceloporus undulatus*), 46
    five-lined skink (*Eumeces fasciatus*), 47
Lumbricus terrestris (Earthworm), 73
Lycaena phlaeas (American copper butterfly), 56
Lygaeus kalmi (Small eastern milkweed bug), 62

M

Mantis religiosa (Praying mantis), 71
Marmota monax (Woodchuck), 17
Marsupials, 25
Melanoplus differentialis (Differential grasshopper), 66
Melospiza melodia (Song sparrow), 82
Mephitis mephitis (Striped skunk), 35
Microtus pennsylvanicus (Meadow vole), 29
Millipede (genus – *Julus*), 68
Mimus polyglottos (Northern mockingbird), 76
Mockingbird, northern (*Mimus polyglottos*), 76
Mole,

eastern (*Scalopus aquaticus*), 22
star-nosed (*Condylura cristata*), 22
Molothrus ater (Brown-headed cowbird), 93
Monomorium minimum (Little black ant), 72
Moths,
  cecropia (*Hyalophoria cecropia*), 59
  clear-winged sphinx (*Hemaris thysbe*), 58
  hummingbird *see* clear-winged sphinx
  Isabella tiger (*Isia isabella*), 58
  luna (*Actias luna*), 59
  white-lined sphinx (*Hyles lineata*), 58

Mouse, 7
  deer (*Peromyscus maniculatis*), 26
  field *see* vole, meadow
  harvest,
    eastern – *Reithrodontomys humulis*
    western – *Reithrodontomys megalotis*, 26
  house (*Mus musculus*), 28
  meadow *see* vole, meadow
  white-footed (*Peromyscus leucopus*), 27
Mus musculus (House mouse), 28
Mustela frenata (Long-tailed weasel), 34
Myotis lucifugus (Little brown bat), 23

N

Nerodia sipedon (Northern water snake), 50
Northern walkingstick (*Diapheromera femorata*), 70
Notophthalmus viridescens (Eastern newt), 37
Nuthatch, white-breasted (*Sitta carolinensis*), 86
Nymphalis antiopa (Mourning cloak butterfly), 55

O

Odocoiles virginianus (White-tailed deer), 6, 36
Ophedrys vernalis (Smooth green snake), 51
Ordonata (Dragonflies and damselflies), 60

P

Papilio glaucus (Tiger swallowtail butterfly), 56
Pardosa – genus (Thin-legged wolf spider), 63
Parus atricapillus (Black-capped chickadee), 87
Passer domesticus (House sparrow), 84
Peromyscus,
  leucopus (White-footed mouse), 27
  maniculatus (Deer mouse), 26
Phalangium opilio (Brown daddy-long-legs), 67

Photuris pennsylvanicus (Pennsylvania firefly), 60
Picoides pubescens (Downy woodpecker), 89
Pieris rapae (Cabbage white butterfly), 54

Plethodon,
  cinereus (Red-backed salamander), 38
  glutinosus (Slimy salamander), 38
Popilla japonica (Japanese beetle), 70
Porcupine (*Erethizon dorsatum*), 29
Possum *see* Virginia opossum
Praying mantis (*Mantis religiosa*), 71
Procyon lotor (Raccoon), 6, 10, 33
Pseudacris triseriata (Chorus frog), 43
Pterophylla camellifolia (True katydid), 65

Q

Quisalus quiscula (Common grackle), 91

R

Rabbit, Eastern cottontail (*Sylvilagus floridanus*), 19
Raccoon (*Procyon lotor*), 6, 10, 33
Rana,
  catesbeiana (bullfrog), 40
  clamitans (Green frog), 40
  palustrus (Pickerel frog), 41
  sylvatica (Wood frog), 41
Red admiral, 8
Reithrodontomys,
  humulis (eastern – Harvest mouse), 26
  megalotis (western – Harvest mouse), 26
Robin, American (*Turdus migratorius*), 74

S

Salamanders, 37–9
  eastern newt (*Notophthalmus viridescens*), 37
  red-backed (*Plethodon cinereus*), 38
  slimy (*Plethodon glutinosus*), 38
  spotted (*Ambystoma maculatum*), 39
  tiger (*Ambystoma tigrinum*), 39
Scalopus aquaticus (Eastern mole), 22
Sceloporus undulatus (Eastern fence lizard), 46
Schistocerca americana (American bird grasshopper), 66
Sciurus,
  carolinensis (Gray squirrel), 13
  niger (Fox squirrel), 14
Seattle, Chief, 6
Shrew,
  cinereous shrew *see* masked shrew
  masked (*Sorex cinereus*), 21

short-tailed (*Blarina brevicauda*), 21
Sitta carolinensis (White-breasted nuthatch), 86
Skink, five-lined (*Eumeces fasciatus*), 47
Skunk,
  eastern spotted (*Spilogale putorius*), 35
  striped (*Mephitis mephitis*), 35
Snake,
  common garter (*Thamnophis sirtalis*), 50
  hognose,
    eastern – *Heterodon platyrhinos*
    western – *Heterodon nasicuse*, 49
  milk (*Lampropeltis triangulum*), 49
  northern water (*Nerodia sipedon*), 50
  rat (*Elaphe obsoleta*), 48
  ringneck (*Diadophis punctatus*), 48
  smooth green (*Opheodrys vernalis*), 51
Sorex cinereus (Masked shrew, 21
Sparrow,
  house (*Passer domesticus*), 84
  song (*Melospiza melodia*), 82
  white-throated (*Zonotrichia albicollis*), 77
Spermophilus tridecemlineatuus (Thirteen-lined ground squirrel), 18
Spider,
  garden (*Araneus diadematus*), 64
  grass (*Genus agelenopsis*), 63
  orb-weaving (*Araneidae*), 64
  thin-legged wolf (*Genus pardosa*), 63
Spilogale putorius (Eastern spotted skunk), 35
Squirrel,
  flying,
    northern – *Glaucomys sabrinus*
    southern – *Glaucomys volanus*, 16
  fox (*Sciurus niger*), 14
  gray (*Tamiasciurus hudsonicus*), 15
  thirteen-lined ground (*Spermophilus tridecemlineatuus*), 18
Starling, European (*Sturnus vulgaris*), 75
Sturnus vulgaris (European starling), 75
Swallow, barn (*Hirundo rustica*), 83
Sylvilagus floridanus (Eastern cottontail), 19

T

Tachinid flies (*Tachinidae*), 61, 70
Tachinidae (Tachinid flies), 61
Tamias striatus (Eastern chipmunk), 12
Tamiasciurus hudsonicus (Red squirrel), 15
Terrapene,
  carolina (Box turtle – eastern), 45
  ornata (Box turtle – western), 45
Thamnophis sirtalis (Common garter snake), 50
Toad, American (*Bufo americanus*), 44
Troglodytes aedon (House wren), 90

Turdus migratorius (American robin), 74
Turtle,
  box,
    eastern – *Terrapene carolina*
    western – *Terrapene ornata*, 6, 45
  painted (*Chrysemys picta*), 45

U

Ursus americanus Black bear), 32

V

Vespula,
  maculata (bald-faced hornet), 52
  maculifrons (eastern – Yellow jacket), 52
  pennyslvanica (western – Yellow jacket), 52
Virginia opossum (*Didelphis virginiana*), 25
Vole, meadow (*Microtus pennsylvanicus*), 27
Vulpes vulpes (Red fox), 31

W

Weasel, long-tailed (*Mustela frenata*), 34
Whistle pig *see* Woodchuck
Wild canary *see* Goldfinch, American
Woodchuck (*Marmota monax*), 7, 17, 18
Woodpecker, downy (*Picoides pubescens*), 89
Wren, house (*Troglodytes aedon*), 90

Y

Yellow jacket, eastern – *Vespula maculifrons*
  western – *Vespula pennsylvanica*, 52

Z

Zenaida macroura (Mourning dove), 88
Zonotrichia albicollis (White-throated sparrow), 77

Credits

Key: *t* = top; *b* = bottom; *l* = left; *r* = right; *c* = center; *m* = main picture.

Jacket credits
Front jacket: © Leonard Lee Rue III (*m*); © Dwight R Kuhn (*cl, cr*); © Gregory K Scott (*l*); © Joe McDonald (*r*).
Back jacket: © Photri/F G Irwin.

© **Thomas C Boyden:** pages 54 *l*, 57. © **Richard Day:** page 14. © **Tom Edwards:** page 8 *t*. © **Dede Gilman:** page 67. © **Kerry A Grim:** pages 27 *l*, 38 *l*, 48 *l*. © **Dwight R Kuhn:** pages 22 *l*, 24, 53 *l*, 54 *r*, 59 *l r*, 66 *l*, 68 *r*, 80. © **McCracken Photographers:** page 70 *l*. © **Joe McDonald:** pages 6 *b*, 8 *b*, 13, 17, 20, 21 *r*, 23, 25, 26 *r*, 27 *r*, 29, 30, 37, 38 *r*, 39 *l*, 40 *l r*, 42 *l r*, 43 *r*, 45 *l r*, 46 *l r*, 47, 48 *r*, 49 *r*, 50 *l*, 51. © **Rick Marsi:** pages 7, 39 *r*. © **Maslowski Photo:** pages 84, 87, 90, 93. **Photri:** pages 33, 35 *l* (© Blakesley). © **Darren Plante:** page 66 *r*. © **J H Robinson:** pages 62 *r*, 63 *l*, 65 *l*, 68 *l*, 72 *r*. © **Leonard Lee Rue III:** pages 6 *t*, 16, 21 *l*, 22 *r*, 28, 32, 44, 49 *l*. © **Len Rue Jr:** pages 36, 41 *r*. **Marcus Schneck:** pages 19, 41 *l*, 50 *r*, 53 *r*, 60 *r*, 73, 78, 91. © **Gregory K Scott:** pages 9, 12, 15, 18, 26 *l*, 31, 34, 35 *r*, 43 *l*, 52 *l r*, 55 *l*, 56 *r*, 58 *r*, 60 *l*, 61 *l*, 62 *l*, 63 *r*, 64, 65 *r*, 69, 70 *r*, 72 *l*, 74, 75, 77, 79, 81, 82, 83, 85, 86, 88, 89, 92. **Unicorn Stock Photos:** pages 58 *l* (© Martha McBride), 61 *r* (© A Gurmankin). © **Scott Weidensaul:** pages 55 *r*, 56 *l*, 76. © **Eric A Wessman:** page 71.